The Exultant Ark

The Exultant Ark

A Pictorial Tour of Animal Pleasure

JONATHAN BALCOMBE

UNIVERSITY OF CALIFORNIA PRESS Berkeley Los Angeles London

University of California Press, one of the most distinguished university presses in the United States, enriches lives around the world by advancing scholarship in the humanities, social sciences, and natural sciences. Its activities are supported by the UC Press Foundation and by philanthropic contributions from individuals and institutions. For more information, visit www.ucpress.edu.

University of California Press
Berkeley and Los Angeles, California

University of California Press, Ltd.
London, England

Prepress by IO Color
Printed and bound by CS Graphics, Pte. Ltd.
Text: Chapparal Pro
Display: Unit
Design and composition by Lia Tjandra
Index by Thérèse Shere

Title page image: Northern Hawk Owl (*Surnia ulula*), Stoney Creek, Ontario, Canada. Photo: Raymond Barlow.

Library of Congress Cataloging-in-Publication Data

Balcombe, Jonathan P.
The exultant ark : a pictorial tour of animal pleasure / Jonathan Balcombe.
p. cm.
Includes bibliographical references and index.
ISBN 978-0-520-26024-5 (cloth : alk. paper)
1. Emotions in animals. 2. Emotions in animals—Pictorial works. 3. Animal behavior. 4. Animal behavior—Pictorial works. 5. Pleasure. 6. Pleasure—Pictorial works. I. Title.
QL785.27.B347 2011
591.5022'2—dc22
2010043747

Printed in Singapore

20 19 18 17 16 15 14 13 12 11
10 9 8 7 6 5 4 3 2 1

The paper used in this publication meets the minimum requirements of ANSI/NISO Z39.48-1992 (R 1997) (*Permanence of Paper*).

To all the animals caught up in the joys and the travails of life.

CONTENTS

Acknowledgments ...ix   Introduction ...1

Play

20

Food

40

Touch

62

Courtship and Sex

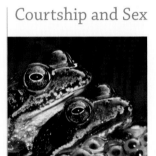

82

Love

Comfort

Companionship

Other Pleasures

102

122

142

164

Conclusion: Implications of Animal Pleasure ...185

Notes ...197   References ...200   Index ...207

ACKNOWLEDGMENTS

This book would be a mere shell without the photographs. I am deeply grateful to all of the photographers for their talent, their patience, and their skill in noticing what animals are doing and for documenting aspects of their lives commonly overlooked. This goes for both those whose works appear in this volume and the many whose fine images had to be excluded because there just wasn't room.

Thank you to those who provided comments and suggestions on the text, helped find images, or assisted with research: Marilyn and Emily Balcombe, Marc Bekoff, Jack Bradbury, Ben Elley, Karen Ferro, Stacy Frank, Peg Lau Hee, Marie Levine, Lori Marino, Sy Montgomery, Cynthia Moss, Shalese Murray, Roni Ostreicher, Kathryn Pasternak, Jaymi Peyton, Fransje van Riel, Erich Ritter, Ken Shapiro, Martin Stephens, Lorel Ward, Amotz Zahavi, and two anonymous reviewers for University of California Press.

Heartfelt thanks to my editor at the press, Jenny Wapner, who also helped shape the text and shepherded the book through a sometimes arduous and difficult peer review process. My views about animals and our relationship to them do not sit comfortably with many of my scientist colleagues. I commend the Regents of the University of California for recognizing the possibility that animals might deserve a better seat at the table than society currently grants them.

Thanks to the rest of the University of California team for their proficiency and professionalism in overseeing the considerable logistics required to bring together a project like this: Lynn Meinhardt, Hannah Love, Dore Brown, Juliana Froggatt, and Lia Tjandra. And thanks to my agent, Sheila Ableman, for always being an advocate for me and my ideas.

I am grateful to the Physicians Committee for Responsible Medicine, which arranged a photo competition for this book; the winning entry appears on pages 152–153.

Finally, thanks to my parents, Maureen and Gerry, for your passion and compassion for animals.

INTRODUCTION

Each winter, throngs of crows gather at the shopping plaza near my home in suburban Maryland. These birds have a lust for life; on the most bitterly cold mornings, I walk briskly to the bus stop in multilayered winter clothing while the crows, garbed in the same outfits they wear in summer, flap about and cavort boisterously as if the chill were nothing. Some mornings two hundred or more of the lustrous black birds are drawn to the Dumpster at one end of the parking lot. They loiter there like spectators at a sporting event, keeping a close eye on the action taking place below. The boldest individuals venture into the Dumpster to explore and tug at promising morsels while others perch vigilantly along rooftops and the branches of naked trees nearby. Often an aerial chase ensues when a bird grabs a hunk of food in her bill and thus becomes the target of others. Sometimes ten or more birds take up the pursuit. Occasionally a ring-billed gull joins in. The target crow tries to elude the pack with dramatic swoops, rolling swerves, and straight sprints. These chases can last many minutes. Schooled in evolutionary biology, I had assumed such behavior was part of the earnest struggle of life. I figured that the pursuers were hungry and desperate for a piece of the spoils. But I've watched such chases through my binoculars, and I've noticed that sometimes the bird being pursued has no food in her bill. One chase lasted eleven minutes and was still going on when my bus arrived. I began to wonder if these aerial skirmishes could be a game.

If the crows are indeed playing a game, then perhaps enjoyment is their primary motivation. As we'll see in the first chapter, many animals engage in play. (For simplicity and convenience, I use the term *animal* to mean a creature other than the human animal.) There are good evolutionary reasons for play behavior, such as developing strength, practicing important survival skills, and learning the social rules of one's species. But animals do not study evolution, and it is not likely that crows consciously play for evolutionary benefit any more than we humans do. They play because it is fun. Understanding animal pleasure requires recognizing that animals feel things both physically and emotionally; they have wants and desires as well as simple biological needs.

Chances are that you have witnessed expressions of animal pleasure. If you've lived with dogs or cats you may have noticed how most of them enjoy being stroked, scratched, or rubbed. Indeed, that we refer to them as pets attests to their love of touch as well as our pleasure in touching them. One of our cats, Megan, adores a belly rub, which she solicits by making a distinctive chirruping sound and flopping onto her side or back. It's a cat billboard that says: *Here is my belly. Please start rubbing.* How can one resist an invitation like that? As I rub, she stretches out to her full length, flopping from one side to the other and purring loudly. If I pick up the cat brush and thump it on the floor, Megan doesn't just walk over—she comes running. In the morning as I comb my hair, she hops up onto the vanity and waits expectantly for her turn.

Our nonhuman companions show a range of emotions and moods. When I was a boy, my family had a small mongrel dog named Begs. It was a treat to watch him after his bimonthly bath. He appeared to dislike the actual bathing process, looking forlorn and with his tail slack as squeegees oozed suds down his sides. But ten minutes later, after being rinsed and dried, he would explode with glee, sprinting about the house, veering in and out of rooms, and bouncing off laps and furniture. Perhaps he was celebrating liberation from the ordeal, or maybe he was simply feeling fantastic, as many of us do after a bath or shower.

These examples are anecdotes, which tend to be less convincing than carefully designed scientific studies that control variables and have repeated samples. An anecdote is a single, chance event. As such, it is more open to different interpretations and explanations. For instance, if someone observed one or two vervet monkeys making an unusual vocalization when an eagle flew overhead, followed by all the other vervet monkeys rushing to the lower branches of any available tree, they might suppose that vervet monkeys have a special call for an eagle predator. But perhaps the vervets were reacting only to the sight of the eagle and not to the calls. Or maybe they rushed to the tree because a vervet had seen a python in the grass, or because the call was made in response to the red shirt the human researcher happened to be wearing that day. These alternative explanations can be eliminated only by repeated observation, usually involving controlled experiments. If, for instance, the vervet monkeys are observed to make and respond to a similar call whenever an eagle flies by, then the case for its being a specific call denoting an eagle becomes stronger. But the best evidence comes from playing the monkeys a prerecorded call, which has been shown to cause vervet monkeys to make a dash for lower tree branches even when there is no eagle present. This experiment shows quite convincingly that the monkeys have a special call for *eagle*, or at least for *aerial predator*.[1]

As yet, there are few scientific studies on animal pleasure. Some of these, however, are very good, and I mention them in the pages ahead. There are also solid arguments for animal pleasure, and plenty of observations to support them. But before I discuss these findings further, I would like to provide some background to the controversial nature of this important topic and its neglect in the scientific community.

I should first clarify my use of three related words that appear often in this book: *experience*, *feelings*, and *pleasure*. The verb *experience* means to observe and participate in an event. It implies that the animal who does it has conscious awareness and may anticipate future events and remember past ones. For example, some fishes monitor the behavior of others and selectively team up with more reliable and skilled individuals for foraging or predator inspection forays.[2] The ability to recognize other individuals and to discriminate among them based on past observations bespeaks a conscious experience of events, so I conclude that these fishes have some level of awareness. I believe that existing evidence, and common sense, supports the conclusion that all vertebrate animals are sentient—they can feel pain and pleasure—and have experiences, and I discuss this more below. Incidentally, in referring to a sentient animal, I prefer the subjective *who* to the objective *that* because we are talking about individual beings with unique identities.

When I refer to *feelings*, I'm making a general statement about an animal's physical or emotional state. In the physical domain, an animal may be hot or cold, tired or well rested. In the emotional domain, an animal may, for example, feel fear, excitement, boredom, or optimism. By *pleasure*, I mean any form of positive experience. Pleasure can be physical (the feel of air-conditioning on a hot day), psychological (pride at receiving an award), or both (the responses aroused by a kiss).

THE CONTROVERSY OVER ANIMAL FEELINGS

I argue throughout this book that pleasure is central to animal existence. Believe it or not, this is a controversial claim. There has been very little discussion of animal pleasure by biologists. Consult the index of an animal behavior textbook and you almost certainly will not find the word *pleasure*. Indeed, there is little serious discussion of pleasure in humans, let alone nonhumans. Pain, yes; pleasure, no. I know of twenty-three scholarly English-language journals dedicated to the study of pain. By contrast, with the exception of an obscure and now-defunct publication called the *Journal of Happiness Studies*, there are no counterparts devoted to exploring pleasure.

Why do we shun pleasure? Part of the reason is that science, by and large, has held and continues to hold a narrow perspective in its scholarly interpretation of animal existence. Published studies of animal behavior are presented almost exclusively in an ultimate, evolutionary context, without discussion of the animals' more proximate mental and emotional experiences. By *ultimate*, I mean the big-picture causes we may not be thinking about, like the adaptive bases for play (such as developing physical strength and social skills); by *proximate*, I'm referring to a more immediate basis for our behavior, such as the desire for fun as a motivator for play. Scientists are more comfortable with ultimate explanations for behavior because these don't require ascribing to animals experiences that are very hard, if not impossible, to be certain of. For example, the researchers who conducted a study showing that Norway rats and golden hamsters preferred novel foods after several days of eating a single food concluded that the subjects wanted to avoid either becoming overdependent on a potentially short-lived food source or risking a micronutrient deficiency.[3] If the scientists had also suggested that the rats and hamsters had grown tired of the same old fare and enjoyed something different, they might have compromised their chances of getting their paper published. Please note that these ultimate and proximate explanations are not in conflict. There is little doubt that rats, like humans, gain a nutritional benefit by varying their diets, but that doesn't preclude their enjoying it. On the contrary, the adaptive benefits of dietary variance have likely driven the evolution of an associated pleasure.

Scientists tend to shy away from making assumptions, and anthropomorphism—the attribution of human characteristics to nonhumans—is generally frowned upon. We inescapably anthropomorphize because we can't help it—we are anthropoid apes, and we cannot know absolutely what other animals are feeling. But we can make reasonable conjectures backed by good science. As Gordon Burghardt, an ethologist (one who studies animal behavior) at the University of Tennessee, explains, we can be "critically anthropomorphic" by having a firm knowledge of the life history, behavior, and ecology of the animals we study.[4] Marc Bekoff, another ethologist, advocates for "biocentric anthropomorphism," by which we try to consider the animal's point of view (hence *biocentric*) and not just our own, anthropocentric one.[5] The key point here is that there is nothing wrong with interpreting animals' behavior in light of our own known experiences provided we are judicious about it. As we move further away from humans (the referent animal whose feelings we know from personal experience), we need to be more cautious. Because we diverged from chimpanzees more recently in evolutionary time than from chinchillas, we can be more confident in drawing parallels between a chimp's behavior and our own. In the face of current knowledge, it seems a bigger assumption that animals are unconscious and unfeeling than that they are sentient, emotional, and aware, and the primatologist Frans de Waal has proposed the term *anthropodenial* "for the *a priori* rejection of shared characteristics between humans and animals."[6]

The study of animal experience has not always been neglected. In 1872 Charles Darwin devoted an entire book to the subject, *The Expression of the Emotions in Man and Animals*. He took a comparative approach, presenting observations and anecdotes to argue for the emotional experiences of nonhumans based on correspondences between human and nonhuman subjects. Then, beginning early in the twentieth century, science fell under the spell of a way of thinking that rejected the study of animal experience. Bolstered in part by Ivan Pavlov's celebrated studies of conditioned reflexes in dogs, behaviorism emerged as the prevailing scientific paradigm. Its adherents hold that behavior can and should be studied without appeal to motives, intentions, or internal states—that is, to thoughts or feelings. This view held sway until the 1970s, when an American ethologist named Donald Griffin published *The Question of Animal Awareness* (1976), the first of three books and numerous articles he wrote on animal thinking and consciousness (see references). Having already codiscovered bats' echolocation (in 1938) and written several books, Griffin was a respected scientist, and today he is widely credited with making the study of animal cognition an active and respectable discipline.

Why has the scientific community taken so long to recognize our fellow animals as thinking, feeling beings? The main reason is that feelings are experienced privately, which makes them hard to test. The sensory experiences of an individual—including another human—cannot, literally, be felt by another. Furthermore, to the degree that animals use communication signals (sounds, smells, body language, facial expressions, etc.) in ways whose meaning we may not understand, and most notably because they don't use our highly developed linguistic channels, we are further disadvantaged in trying to interpret what and how they are feeling.

For instance, alternative hypotheses can be offered to explain the behavior of the crows I described above. One possibility is that their aerial chases involve the expression or establishment of a dominance hierarchy. Such hierarchies have been reported in crows; in a European study, male carrion crows tended to prevail over females, adults over juveniles, and larger over smaller individuals for access to a clumped food source (such as the Dumpster in my neighborhood).[7] The question of whether the aerial chases I saw were displays of dominance could be tested (with difficulty) by marking birds so that individuals could be recognized. If chases corresponded to a hierarchical pattern, then the dominance hypothesis would gain support. If, however, there was little or no pattern to who chased whom, we might seek a different explanation.

Another possibility is that the crows are merely practicing the art of piracy—the behavior of wresting a tidbit from another by the use of force or intimidation. Gulls and frigatebirds are among the bird species best known for aerial piracy; they will pursue and harry another bird who is carrying food in an attempt to compel him or her to drop the prized object. If the crows are boning up on skills needed for effective piracy, then these chases qualify as play behavior. It should be noted that neither the dominance nor the piracy practice hypothesis precludes the possibility that the crows enjoy this activity. How they feel is very difficult to test scientifically, but we can gain insight through study and observation. Animals are not closed books, and their feelings are not unsolvable mysteries. We can discover a great deal about animals' experiences by following, for instance, Charles Darwin's method in *The Expression of the Emotions in Man and Animals*. Many animals—particularly species that live in social groups—are very expressive. Provided we have a good basic knowledge of the ecology and behavioral repertoire of a particular species, we can learn from observing and studying its members.

Few scientists today are willing to deny that vertebrate animals are sentient—capable of feeling pains and pleasures—though there are legitimate discussions of where in the animal kingdom we ought to draw the line on what animals are and are not sentient (more on this below). The fortunate consequence of a more open-minded era is that science is applying more creative methods to the study of animal cognition, awareness, and emotion. Throughout this book I present examples of studies that scientists are now conducting to better understand what and how animals think, and their emotional feelings—particularly as they may relate to pleasurable experiences. Allowing animals to make choices among a series of putatively desirable options—say, types of food—is a useful tool for assessing what they like. We can also measure physiological changes, such as the release of endorphins and other pleasure-associated compounds from the brain or changes in heart rate or blood pressure. In all of these instances, the human response provides a helpful reference point, provided we bear in mind that our own preferences and reactions are not always reliable predictors of those of other species.

THE CASE FOR ANIMAL PLEASURE

For all of the progress made by science regarding animal feelings, the topic of pleasure remains nascent and largely neglected in scientific discourse. Until animal pleasure is widely and academically accepted, a detailed defense of its presence is warranted. Here, then, I present three primary arguments and hypotheses supporting the case that animals feel pleasure.

First, I propose that pleasure is adaptive. In a "carrot and stick" world, pleasure is nature's carrot. Pain discourages animals from doing things that risk harm or death, which are not good outcomes in the evolutionary stakes. Pleasure, on the other hand, is nature's way of improving survival and reproductive output. Pleasure evolves in sentient organisms as a consequence of behaviors (e.g., feeding, mating) that generate "good" outcomes (e.g., sustenance, offspring) and/or as a motivation to engage in these behaviors based on past rewarding experience.

About a billion years ago, when organisms first began moving about, an adaptive premium was placed on sensory systems. The ability to perceive and respond actively to their environment allowed these early animals to orient themselves toward favorable things such as food sources and others of their kind (especially useful for sexual reproduction). Senses also helped creatures to avoid aversive things such as solid objects that could cause injury, or other animals that might eat them. Through natural selection, organisms better able to sense and orient accordingly developed an advantage over organisms less endowed, and they therefore tended to leave more offspring, which carried their genes for perceptual skill. Thus, the evolution of mobility was a key step toward the eventual evolution of feelings. Just as pain or discomfort motivates the sentient organism to move away from something aversive, so too does pleasure benefit by rewarding the individual for performing behaviors that promote survival and procreation.[8]

The second argument for animals' experience of pleasure is that we know pleasure exists in at least one animal: the human. Even though, as I stated earlier, sensations are private (in that no other individual can actually feel another's feelings), we readily accept that humans are sentient because each of us has our own experience to inform us. That human languages contain rich vocabularies for describing good feelings—happiness, delight, surprise, anticipation, pride, satisfaction, joy, elation, ecstasy, thrill, euphoria, exultation, jubilation, excitement, rapture, fulfillment, gratification, and comfort, among others—attests to the diversity of both physical and emotional pleasures that can be felt by humans. Our knowledge and acceptance of these sensory phenomena in one species provide a firm foundation for their existence in others. With humans as our necessary base, we can radiate out phylogenetically to other animals and ask to what degree we can test similar experiences there.

We should not assume that the range of pleasurable experiences felt by animals is delimited by those of humans. It is possible that our complex social networks and sophisticated languages have given rise to certain subtle emotions—satisfaction and gloating, for instance—that are either absent from or poorly developed in other animals. But animals, having evolved in diverse environments that present novel challenges and opportunities, have developed a great wealth of sensory capacities and perceptual skill sets. Many of these lie beyond our sensory bandwidths. For example, bats and whales orient themselves using echolocation; some fishes communicate with pulses of electricity; birds and fishes can see a broader spectrum of colors thanks to the four types of photopigments in their eyes (most humans have three); and sea lions can track fishes in unlit waters by detecting turbulence trails with their whiskers.[9] Some species can tune in to the earth's magnetic field to help navigate, and there is now evidence, based on a study of garden warblers, that birds perceive it visually.[10] To the degree that sensory systems give rise to sentience and sentience to feelings, animals may experience forms of pleasure inaccessible to humans. These examples do not explicitly involve pleasure, but they do illustrate the potential for pleasures unknown to us.

A third argument for pleasure in animals is that they are equipped to feel it. All members of the vertebrate kingdom—mammals, birds, reptiles, amphibians, and fishes—share the same basic body plan: skeletal, muscular, nervous, respiratory, circulatory, digestive, excretory, endocrine, reproductive, and sensory systems. The last gives them five basic senses—sight, smell, hearing, touch, and taste—which serve as the interface between an animal's nervous system and its surroundings. Equipped with the ability to detect and avoid unpleasant stimuli and to seek rewards, animals have the raw materials on which natural selection can act to favor pleasure and pain.

In addition to these structural and sensory features, we share with animals many physiological and biochemical responses to sensory events. Pleasure and reward are generated by brain circuits that are largely similar in humans and other vertebrates. When humans experience a painful or pleasurable sensation, our brains and our glands secrete chemicals that help us respond appropriately. For instance, our positive feelings are mediated by such compounds as dopamine, seratonin, and oxytocin. Two brain structures—the amygdala and the hypothalamus—play an important role in human emotions. Many animals, especially mammals, possess these same neurological structures and brain chemicals. Imaging technologies such as PET and MRI provide further evidence that animals experience emotions as we do.[11]

Another aspect to this argument, based on animals' physical and physiological assets, is that they can feel pain and can suffer. Curiously, animals' capacity for pain—while no less physically private than their capacity for pleasure—is uncontroversial. Pain in humans and animals has been extensively studied and is well established.

Studies have shown that the processes of perceiving unpleasant or painful stimuli and relaying them to parts of the brain that register pain are practically identical among the different mammals that have been examined. Measuring pain perception at the brain level is more accurate than observing the external behavior of animals, which can be misleading. For example, some animals remain stoic even when they're in intense pain; this is an adaptive response that developed to avoid showing weakness or vulnerability—very useful when one is being watched by a potential predator! With regard to experiencing pain, there are no unequivocally more or less sentient species, at least among mammals.[12] Might the same be said of pleasure?

Most animals in pain express it behaviorally. They shriek or bellow, they avoid and retreat from sources of pain, and they flinch, limp, and protect the injured part. These responses are all compellingly consistent with the sentient experience of pain. Nature also provides many indirect clues to animal pain. Plants, for example, have exploited animals' capacity for pain and discomfort with the evolution of thorns and bitter-tasting chemical compounds in their tissues. Similarly, many animals have well-developed spines, stingers, horns, and tusks, which, like thorns, can inflict pain and signal *don't touch*.

That injured animals experience pain is reinforced by scientific studies. Laboratory research has shown repeatedly that injured rats, for example, will favor the bitter taste of water that contains a pain-relieving drug over unadulterated water.[13] Commercially raised chickens often experience leg problems associated with selective breeding, growth, and husbandry and can develop highly abnormal gaits or become completely unable to walk. In this condition, lame birds soon learn to eat from a presented source of food that has been treated with painkillers rather than from a source of similar but unadulterated food. Furthermore, as the severity of lameness increases, lame birds consume a greater proportion of the drugged food.[14]

Evidence is now emerging to suggest that some invertebrates also experience pain. In studies conducted at Queen's University Belfast, prawns (close relatives of shrimps) who had an antenna subjected to mild acid or pinching spent a prolonged period of time grooming the assaulted antenna or rubbing it against the side of the aquarium. Because these responses were not brief and reflexive and because they were inhibited by a pain-relieving drug (benzocaine), the researchers concluded that the prawns were probably experiencing pain.[15] Other research has presented evidence for the experience of pain in crabs. There is also evidence for sentience, including emotional states, in octopuses.[16] These observations raise the intriguing possibility that some invertebrate animals may also feel pleasure. Along these lines, studies of captive crayfish found that they prefer to swim in a quadrant of a test arena into which solutions of cocaine or amphetamine are infused.[17]

Although these arguments create a convincing foundation for animal pleasure—that it may benefit survival, that if humans experience pleasure, other animals are likely to as well, and that ani-

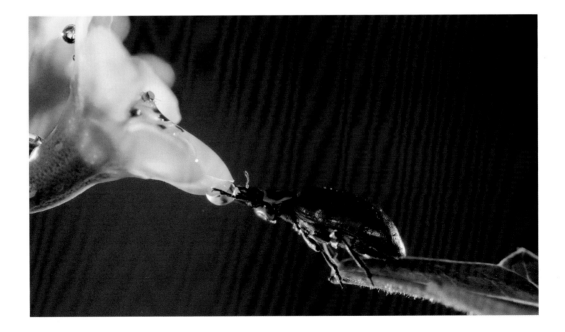

LADYBIRD BEETLE (Coccinellidae sp.), Alice, Texas. We may never know for certain whether insects can have conscious experiences. But if they can, many aspects of their lives should bring them rewards. Photo: Scott Linstead.

mals share with humans the physical and physiological equipment to feel bad (and by analogy, good) things—they support the idea without actually demonstrating it. They do not prove outright that animal pleasure exists. However, I cannot prove outright that even *you* feel pain or pleasure; this is the privacy of sensory experience. But while solipsism—the view that the self is the only thing that can be known to be conscious—is logically incontrovertible, it would nevertheless be extremely dogmatic and cynical to act as if it were true. And based on the evidence science has so far marshalled for animal sentience—bolstered by our own observations—it stretches belief that animals could be insentient.

The remainder of this book is primarily devoted to direct and indirect evidence for animal pleasure—that is, studies and observations (and, of course, photographs) that illustrate the finer moments in an animal's life. Let me reiterate that the study of animal pleasure, or what I have called *hedonic ethology*, is still an infant science. Much more scientific evidence needs to be compiled before the importance of pleasure in animals' lives gains wide recognition. That's one of the reasons I'm so pleased to be writing this book.

THE RICHNESS OF PLEASURE

Today evidence is rapidly accumulating that life for animals holds great potential for joy. In a freshwater spring in Kenya, hippopotami drift blissfully, splaying their toes, spreading their legs, and opening their mouths as fishes of various species provide a spa treatment by nibbling away bits of food, sloughing skin, and parasites. In a laboratory in Ohio, young rats come running to receive tickles from the fingers of a trusted human, uttering ultrasonic chirps linked to the origins of human laughter. In Montreal, captive iguanas shun the bland reptile chow beneath their heated perch and venture out to retrieve gourmet hunks of fresh lettuce at the other end of their terrarium.[18] I'll share more on these studies in later chapters.

Scientists who study positive experiences in animals are still few and far between, but one of them is the physiologist Michel Cabanac, at Laval University in Quebec. Cabanac coined the term *alliesthesia* (from the Greek for "other-perception") to describe the phenomenon whereby an identical stimulus may be perceived as pleasant or unpleasant depending on the physiological state of the subject. For example, when Cabanac had human subjects dip their hand in a container of cool or cold water, they reported the experience as pleasant if they were feeling hot (e.g., after emerging from a sauna) and unpleasant if they were feeling cold (e.g., after emerging from a freezer).[19] Animals too show alliesthesia, which also applies to tastes (pleasant when hungry, unpleasant when full), though not, for instance, to sounds and lights. Rats reverse their preference from sweet to pure water when they are dehydrated, a change that shows positive alliesthesia for water.[20] Similarly, nature rewards a cold animal who finds warmth, and vice versa. For alliesthesia to work, an animal need only be able to experience surroundings as pleasant or unpleasant and have the physical means of moving to or otherwise selecting the preferred stimulus.

As Darwin showed, we humans are just one of many wonderful and unique expressions of nature: our differences from other mammals (at least) are in degree, not kind. Crucially, when it comes to sentience, humans may not always be the most endowed. I am not aware of any scientific attempts to compare the intensity of pleasurable or painful feelings between humans and other animals, but some scientists, such as the American neuroscientists Jeffrey Burgdorf and Jaak Panksepp, think that other life-forms may experience certain feelings more intensely than humans do.[21] When my cat Megan receives her belly rubs, she seems totally absorbed in the moment—enwrapped in a pleasure whose pureness may be more difficult to attain for us whose minds become easily preoccupied with our thoughts. On the flip side, other species might suffer pain or distress more than us if they are not able to rationalize the source of their suffering. In *The Unheeded Cry*, the American bioethicist Bernard Rollin suggests that animals with a reduced concept of time may not look forward to or anticipate the cessation of pain: "If they are in pain, their whole universe is pain; there is no horizon; they are their pain."[22]

We already know that many animals have keener senses than our own. Owls have better night vision and hearing, sharks have stronger chemical perception, and dogs have a better sense of smell. In some cases, animals experience physical sensations that are unknown to us. What might it feel like to orient in flight or to identify different types of insect by listening to one's echoes, as bats do? Or to communicate by means of vibrations, as many burrowing animals do, or to sense the earth's magnetic field? The German ethologist Jakob von Uexküll coined the term *Umwelt* (German for "environment" or "surrounding world") in the early twentieth century to denote each animal's unique, subjective perceptual world.[23] Knowing exactly what another animal's sensory experiences feel like may be elusive. But we can grasp a great deal from observing other animals' responses to stimuli by using our personal sensory experience as a template.

The beauty of science is that we don't have to resign ourselves to games of pin-the-tail-on-the-donkey in regard to questions of animal sentience, awareness, or cognition. Careful scientific studies—with random samples and control groups—can help us pursue these challenging questions. We can design and conduct studies to discover, for instance, whether a reptile or an amphibian runs a temperature from the emotional reaction to being handled (a reptile does; an amphibian does not), whether fishes recognize other individual fishes (they can), and whether young chimpanzees outperform humans on short-term spatial memory tests (they do, decisively).[24]

PREJUDICES

Advances in our knowledge and understanding of animal sentience are compelling us to reconsider our prejudices toward animals. One such prejudice is the notion that life in the wild is a relentless, earnest struggle. Traditional portrayals of nature tend to emphasize its more dramatic, violent and bloodthirsty moments. Popular phrases such as "nature red in tooth and claw," "eat or be eaten," and "the struggle for survival" reinforce the impression that life for wild creatures is harsh and grim. I consider this a biased and inaccurate perspective.[25] Consider that survival behaviors in themselves can be rewarding: just because reindeer have to migrate a thousand miles to find seasonally available food, or prairie dogs need to dig burrows to avoid predation, doesn't mean they can't take pleasure in these tasks. Goal-directed activities, such as important survival behaviors, are desirable for animals, who need to exert some control over their lives.[26] Finding food is one of the major projects of an animal's life, and many with the misfortune of being confined have been shown to engage in what is called *contrafreeloading:* given the opportunity, penned pigs will root, caged rats will pull a lever, and pigeons in Skinner boxes will peck at a disk to obtain food that is otherwise freely available without having to engage in these activities.[27] Deprived of opportunities to behave as they would in the wild, caged mice may also become neurotic. Their brains grow stunted, and they spend hours performing repetitive, functionless activities such as gnawing on the cage bars, doing somersaults, or digging at the cage corners.[28] Take away life's significance, and you may be taking away a lot of what pleasure derives from.

Also, despite what we've heard of nature's hardships, animals do in fact have leisure time. Many animals meet their survival needs in a fraction of the time available to them. The primatologist Robert Sapolsky estimates that savannah baboons on the Serengeti plains of Kenya take about four hours to feed themselves in a given day.[29] Dikdiks (small African ungulates) spend about six hours of an average day chewing the cud and another three hours moseying about but not feeding.[30] Flight affords many birds the luxury of meeting their energy needs in a fraction of their waking time. Animals may spend part of the remaining time engaged in such activities as grooming and preening, playing, singing (birds), or resting.

Another commonly held prejudice is that animals raised for food are less worthy of our consideration than wild animals and that domesticated animals are mentally dimmer or less sensitive than their wild ancestors. Pigs are often perceived as filthy, chickens as cowardly and stupid, and sheep as passive followers. This book features quite a few farmed animals. I prefer this term to the more common *farm animals* because it doesn't imply, capriciously, that being raised for consumption is their biological purpose; it's not a role they chose but one we've imposed on them. Nor does it reinforce the prejudice that they are different from wild creatures in any relevant way, such as the capacity for pleasure.

In fact, farmed animals have been well studied, and none of the biases we commonly hold against them stand up to scrutiny. Chickens, for example, have a vocabulary of at least thirty different calls. Some, like the large-, medium-, and small-aerial-predator alarm calls, are referential, meaning that the signaler is referring to a specific object in the environment. Studies by Chris Evans and his colleagues at Macquarie University, in Australia, show that a chicken on the receiving end of these calls understands their meaning.[31] Roosters curry favor with a come-hither call that brings a nearby hen running to receive a food tidbit gallantly offered up to her. A rooster who indulges a hen this way may improve his chances of mating with that hen, say, the following week. Note that such reciprocity requires those "stupid" chickens to recognize and remember individuals and to track past accounts. Sometimes a rooster fakes it, calling when there is no tidbit to be had. Studies have shown that a rooster won't risk this if the hen is too close, presumably because she's more likely to detect the ruse.[32] When the hen arrives, the rooster does his usual thing, pointing to where he "saw" the cricket or grasshopper. Insects don't always stay put, so it's plausible that this one may have escaped to safer

ground. Thus, the hen may be deceived into believing that the rooster meant well. As with all deceptions, its success hinges on roosters' being honest most of the time—hens soon learn to ignore those who deceive too often.

As gregarious creatures, sheep might be expected to recognize their flockmates. Studies led by Keith Kendrick at the Babraham Institute in England have shown that sheep indeed have excellent facial recognition skills. Sheep who were removed from their home flock and placed into another for two years recognized up to fifty of their former flockmates in single photographs presented at random among those of nonfamiliar sheep.[33] Sheep also strongly favor a photograph of a just-fed/contented sheep to that of a hungry/stressed sheep. When hungry sheep were presented with two doorways to some fresh food, one of which had the hungry-sheep photo affixed to it and the other of which had the just-fed-sheep photo, they almost always chose whichever door had the just-fed-sheep photo.[34] Sheep make the same distinction between a smiling human face and a frowning one.[35] The emotional responses are evinced by behavioral and physiological changes such as more-relaxed body movements, fewer protest vocalizations, and lowered heart rates and stress hormone levels in response to preferred stimuli.

If some organisms (e.g., mammals) feel things and others (e.g., shrubs) do not, then there must be some imaginary line we can draw between them. Responding to a stimulus in the manner of a plant like the venus's-flytrap can be explained mechanically, in terms of changes in turgor pressure in cells and the like. But unlike plants, many animals have nervous systems and senses. In the panoply of life on earth, where do we draw the line on pleasure? Our prejudices tend to rule against considering invertebrates as sentient, but there is much evidence, such as in those studies of prawns and octopuses mentioned earlier, that we should rethink this. Could it be that ants' preferences for certain foods or habitats reflect positive experiences they've had? I don't know. We should not be too hasty to deny the possibility that they can. Recently, ants have been shown to forage idiosyncratically—individuals consistently use their own unique routes—and also to quickly learn to take a more direct route between two points (e.g., a nest and a food source).[36] I am not troubled by deny-

DOMESTIC CALF (*Bos taurus*), Farm Sanctuary, Orland, California. Whitaker was found on a highway, frail, sickly, and only a few days old. He was taken to a sanctuary, where he now lives among other rescued farmed animals. Here, several days after his arrival, he expresses his natural curiosity. Calves learn what to eat and what to avoid through sensory experience and by observing the preferences of others in their herd. Another photo of Whitaker appears in the chapter on other pleasures. Photo: Connie Pugh.

ing feelings to a paramecium, but I am slightly uneasy about doing so for an ant. Wherever we may decide to draw our imaginary lines, we should draw them in pencil.

I'm not about to make the case that there are any hedonistic amoebas or ecstatic sponges out there. These organisms can respond to stimuli, but their lack of a centralized nervous system suggests little basis for their having feelings. However, there are many examples of abilities and traits that formerly only humans were thought to possess but that have since been demonstrated in nonhumans. These include tool use, manufacture, and modification; self-awareness; self-recognition; culture; episodic memory; semantic communication; deliberate deception; self-restraint; and a sense of fairness.

Despite doubts as to whether they are sentient, I have included some invertebrates in this book. The term *invertebrate*, while anatomically well defined, is not so when it comes to the capacity to feel. The invertebrates represent an enormous diversity of creatures, occupying some thirty-six different animal phyla—compared to just one phylum (chordates) within which all vertebrate animals are classified. Fewer than one in twenty extant animal species has a backbone. It is premature and simplistic to conclude that all invertebrates lack feelings.

Members of the invertebrate mollusk phylum—in particular the cephalopods—show evidence of sentience. Octopuses have the hallmarks of individual personalities, they recognize individual humans, and they show behavioral evidence of play and even mischief.[37] In captivity, two octopus species have been observed engaging in what might be play behaviors with plastic objects (a Lego block and an empty plastic pill bottle) and with bubbles.[38] Octopuses are more likely to perform these playlike behaviors if they have recently been fed, further suggesting that the activity is recreational.[39] There is even anecdotal evidence that an octopus can hold a grudge. One individual squirted water at the Canadian biologist Jennifer Mather over the course of two weeks after she banged on the aquarium lid through which the octopus was trying to escape. Mather also reports that a red octopus regularly squirted water at a visiting scientist/photographer but never at Mather or her lab assistant.[40] Since 1981, the common octopus has been protected by an animal welfare law in England.

There's another reason to include invertebrates in a book on animal pleasure. Their colors, their symmetry, their textures, and their graceful shapes and movements may bring pleasure to another animal: you! In the chapter titled "Other Pleasures" I examine aesthetic beauty as a possible source of pleasure for other animals.

One more group of animals deserves special mention: fishes. They too are diverse, numbering more species—about twenty-seven thousand—than all the other vertebrates (mammals, birds, reptiles, amphibians) combined. They've been around a long time, and they, like all organisms, continue to evolve. The old belief that fishes are thoughtless, unfeeling automatons—while still predominant in our popular culture—is being put to rest by science. Many researchers are now studying thinking behavior in fishes, and in 2006 a scholarly volume appeared with the title *Fish Cognition and Behavior*.[41] A couple of generations earlier, the mere idea of a book on thinking in fish would have been met with harsh skepticism if not ridicule from most biologists. Today we know better.

Here is a small sampling of what contributors to this book report: In captive studies, individual bluegill sunfishes remembered the amount of food they got when swimming with different individual foraging partners. This was presumed to be perceived by the fishes as an indication of foraging success. Over the course of several weeks, bluegills avoided less fruitful partnerships based on this accumulated knowledge.[42] (Other research has demonstrated preferences for schooling with familiar individuals—for example, by seeing which end of an aquarium tank fishes will swim in to be in proximity to other particular fishes in an adjoining chamber[43]—in at least eighteen species.)

Blind Mexican cave fishes use their lateral line organs to detect small differences in water flow patterns as displaced water reflects off objects in their environments. Experiments have shown that using this sense allows them to quickly learn and memorize the locations of landmarks and obstacles in their environments. Once they have developed their mental maps, they swim faster in familiar surroundings.[44] Male sticklebacks steal eggs from neighboring nests, which bolsters their own attractiveness to females.[45] Siamese fighting fishes alter their threat displays depending on who's watching; if there is a female audience, males tone down their aggression and tailor their postures more toward a sexual display.[46] To escape death, frill-fin gobies sometimes must jump from a drying rock pool to a deeper one at low tide. If the tiny fishes are able to memorize the topography of the area beforehand, at a higher tide, they can avoid having to make a leap of faith. And in fact, captive gobies showed a marked improvement in orientation after an overnight opportunity to swim over the pools during an artificial high tide. Removing the gobies from their home tide pool for various periods of time before retesting their jumping ability showed that their memory of familiar pools lasted about forty days.[47]

With evidence like this, it is getting harder to claim that fishes are insentient. But there are still some biologists who deny that fishes can feel anything. James Rose, professor emeritus from the University of Wyoming, argues that fishes cannot feel pain or suffer because they lack the brain sophistication to experience these things consciously.[48] What they experience, Rose claims, is nociception—a physiological response to pain without the actual feeling of pain. But that viewpoint too is getting harder to sustain. Fishes confronted with stimuli known to cause pain in other animals and humans suspend normal behavior and show changes in their brain activity, as well as responding with aversive and physiological reactions.[49] In a study, trout responded to having painful substances (e.g., acetic acid, bee venom) injected into their lips by rocking from side to side, rubbing their lips on the bottom or walls of the aquarium, taking longer to resume feeding than did fishes injected with a painless saline solution, and resuming feeding sooner if treated with a painkiller (morphine).[50] Many fishes give off an "alarm" pheromone when confronted with danger, and naïve fishes can learn to recognize predatory fish and high-risk habitats from the pheromones given off by other fishes.[51]

Fish feelings aren't necessarily restricted to pain and fear. They may also have a pleasurable side. The well-studied mutualism between cleaner fishes and their clients appears to be mediated, in part, by good feelings. Evolutionarily, clients benefit by having parasites and sloughing skin removed, and cleaners get nourishment in return. For the client, at least, the spa treatment appears to be pleasurable. More on this in the chapter on touch.

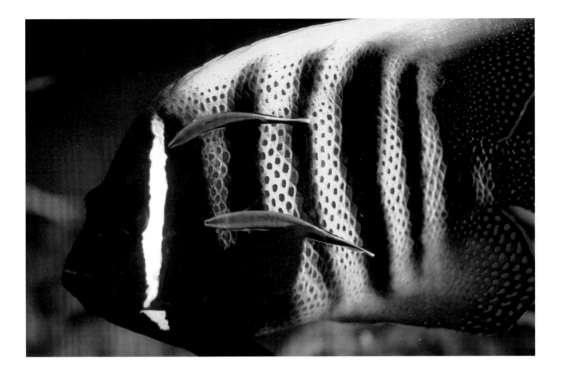

INTERPRETING THE PHOTOGRAPHS IN THIS BOOK

This book is first and foremost intended as a pictorial celebration of animal pleasure. It presents a compilation of photographs depicting animals in mostly pleasurable contexts. The chapters are arranged by topics such as play, touch, food, and sex. Because animals are photogenic and pleasure is a buoyant subject—yet one all too rarely linked to the animals' own experiences—animal pleasure seems an ideal focus for a pictorial book. I hope it works for you as it does for me. The concepts arranged by chapter in this book often overlap, and it was not always easy to assign a photograph to a particular category. For instance, many of the photos in the chapter on touch could have been assigned to companionship or love. I like this, because it illustrates the interplay of different facets of animals' emotional and physical lives. Sea otter companions link arms to keep from drifting away from each other while they sleep afloat on their backs (see page 158). It's a social variation of their habit of tethering themselves to kelp strands for the same purpose. This functionality doesn't preclude feelings of love or attachment toward each other.

SIX-BARRED ANGELFISH (*Pomacanthus sexstriatus*) and BLUE-STREAKED CLEANER WRASSE (*Labroides dimidiatus*), Great Barrier Reef, Australia. New findings reveal fishes to have complex social lives and impressive cognitive skills. Here an angelfish enjoys the ministrations of a pair of cleaner wrasses, who pluck external parasites, algae, and other unwanted debris from their client. It's a reward-based mutualism: the angelfish gets a healthy spa treatment and the wrasses get some nourishment. Many cleaners and their clients form long-term partnerships built on trust and mutual benefit. Photo: Maxi Eckes.

At the outset, there is a key point I must make about the images. While each of the photos is beautiful in itself, some of the animals pictured may not strike you as experiencing pleasure, at least not at first glance. Most animals do not show their feelings in their faces, at least not that we can tell. For example, dolphins experience a range of emotions, and they are notable pleasure seekers, but that fixed "smile" on a dolphin's face (or the fixed expression on a bird's or a fish's face) by itself gives no clue to how she may be feeling. A "smiling" dolphin confined alone in an oceanarium exhibit somewhere is probably unhappy. On the flip side, if a group of dolphins is riding the curl of a large wave breaking toward a beach, we can be fairly confident that they are having a whale of a time. It is the context and behavior—plus familiarity with the species's behavior patterns—that provide our best clues to how an animal is feeling. For this reason, many of the captions in this book are detailed and meticulous. In most cases I consulted the photographer to obtain any information about additional context and behavior that may not be provided by a still image alone.

Another challenge is that a scene's context may not be immediately apparent. When I first viewed Veronika Gaia's photo on this page of a bull elk in the foreground looking away from the photographer toward a herd of cow elk, my impression was that the bull might be feeling sexual excitement as he weighed his prospects. When I consulted Gaia, I learned otherwise. Here's her reply:

> This photograph was an amazing experience for me. It was taken at Rocky Mountain National Park, in Colorado. Two herds of elk with their own bulls started to merge. The females became quite nervous as the two bulls approached and confronted each other. The sound was incredible as they clashed antlers. One bull elk lost his entire harem to the other bull. This image shows him looking at his cows leaving him after the confrontation. He seemed emotionally devastated by his loss. He was bellowing as I took the photo, and the sound affected me for days.

In this case, one bull's gain is another's loss, and what I took to be a pose of pleasurable anticipation in the foreground elk was in fact more likely to be one of defeat, perhaps even bereavement.

Interpreting animal feelings is an imperfect art. There are sometimes many factors to take into account. Anders Nielsen's photo of two yellow-spotted side-necked turtles on the Madre de Dios River in Peru is a case in point. The reptiles are basking in the sun—a favorite pastime of freshwater turtles everywhere. Warming up is important to ectothermic animals ("cold-blooded" is a misleading term, for the sun and their own energy expenditure warm their bodies to temperatures comparable to those of endothermic, or "warm-blooded," animals). I suspect that the sun also *feels* nice on their backs, though there is no scientific evidence (yet) to support or deny this. Turtles will often stretch out their hind legs and splay their webbed hind feet to expose the maximum surface area to the sun. Adaptive? Yes. Pleasant? Probably. When I am out canoeing, I usually try to steer clear enough not to scare basking turtles into the water; I can get a perfectly good look through my binoculars. But occasionally I approach too closely, and I've noticed that on bright, sunny days they will allow me to get closer before slipping from their perches than if it is overcast. Are they more reluctant to leave a warm spot because they don't want to end the pleasure? This is speculation, but questions like this are open to quantitative study.

ELK (*Cervus canadensis*), Rocky Mountain National Park, Colorado. Looks can be deceiving. At first glance, it might appear that the bull elk in the foreground is anticipating his prospects for sexual bounty as he surveys a herd of cow elk. In fact, he has just lost this harem to a competing male. His mournful-sounding bellows haunted the photographer for days. Photo: H. Veronika Gaia.

YELLOW-SPOTTED SIDE-NECKED TURTLE (*Podocnemis unifilis*), Madre de Dios River, Peru. Photographer Anders Nielsen encountered this pair of endangered yellow-spotted side-necked turtles while boating down the Madre de Dios River. Turtles bask in the sun to warm their bodies. This behavior is adaptive, and it probably feels good to them. Notice the butterfly perched on each turtle's nose. They are probably drawn to the salts and other minerals excreted from the turtles' eyes. Photo: Anders Grøndahl Nielsen.

You will probably have noticed some other details in Nielsen's photo. On each turtle's head is perched a butterfly. Nielsen relayed to me that his nature guide told him that the butterflies are drawn to sources of salts and minerals, and that the minerals they seek can be found in the turtle's tears. Also, the turtle on the right is resting a front foot on the one in front. Is this a gesture of friendship? I doubt it. Perhaps it is just more comfortable that way for the turtle basking behind. I have seen turtles crowded onto a prized perching log with two, sometimes three, individuals stacked atop one another. Smaller ones are invariably on top. It's a comical sight reminiscent of a Dr. Seuss book. I suspect these situations are merely a result of opportunism and that they don't involve any particular friendship. But we should not be too hasty to dismiss a social element to turtles' basking behavior. At the very least, these turtles are gregarious and don't mind the presence of other turtles—or butterflies. There are many benefits to social living, including collective vigilance for danger and the exchange of useful information about resources such as food and nesting sites. Turtles also must mate, which requires that they consort socially with others; some species engage in elaborate courtship displays and are selective about whom they mate with.

THE EXPERIENCE OF PLEASURE

The real world runs on experience. For instance, while there may be good adaptive bases for the broad and diverse human cultural practice of adding spices to our foods, I'm not aware that anyone reaches for the oregano or the curry powder with the conscious intent of warding off intestinal microbes. We spice our food because it enhances the taste. Similarly, animals are not mechanical slaves to evolutionary adaptations; they too have experiences. Ray Barlow's photo of a northern hawk owl speaks to me of an animal's conscious awareness. This individual had been flying toward the photographer when something behind caught the owl's attention. Supremely equipped with vision and hearing, owls—hungry ones especially—are on the alert for any sign of nearby activity. Their flexible neck joints allow them to rotate their heads nearly 360 degrees when perched. This species has been known to detect and reach prey buried under a foot of snow. Being aware is about being conscious of one's surroundings and being able to have experiences.

In *Pleasurable Kingdom: Animals and the Nature of Feeling Good* (2006) I presented the argument and the current evidence showing that animals are not merely pain avoiders but also pleasure seekers. That was the first book dedicated to the subject of animal pleasure.

Pleasurable Kingdom presents the subject in words. But animals are photogenic. Their blissful moments are visually charismatic and emotionally contagious. The photographs in the book you are now holding are windows into the inner spaces of other beings. They show animal life as so much more than survival. They depict life's richness and put a more congenial face on nature.

Happy, healthy animals are beautiful to behold. They make us smile, and there's value in that. But pleasure has deeper meaning and significant implications for humankind's relationship with other animals. Pleasure adds intrinsic value to life—that is, value to the individual who feels it regardless of any perceived worth to anyone else. Pleasure seekers have wants, needs, desires, and lives worth living. They can have a good quality of life.

As you leaf through this book, enjoy the pleasure these photos bring you. Bathe in their beauty and soak in their grace. Reflect on the significance of the fact that animals also experience good feelings. And the next time you see a crow or a cat or a lizard, stop and watch. Try to imagine their experience.

NORTHERN HAWK OWL (*Surnia ulula*), Stoney Creek, Ontario, Canada. The posture of this hawk owl—looking over her shoulder as something catches her attention midflight—illustrates the keen awareness animals have of their surroundings. It is with their senses that animals are able not only to survive but, like us, to enjoy the experience. Photo: Raymond Barlow.

AMERICAN ROBIN (*Turdus migratorius*), Bolinas Ridge, Mount Tamalpais, California. These three robins were among a huge flock flying up the side of Bolinas Ridge in Marin County. The photographer reports that it was an extremely windy morning, giving the impression that the birds were flying in slow motion. No doubt it was hard work for them. But this image shows something else: the ineffable quality of freedom. Photo: Trish Carney.

Play

O f all the behaviors animals engage in, play is the least controversial in its suggestion of feelings of pleasure. Animal play has an unmistakable quality. The whole comportment of the participants exudes joie de vivre. Playing has an important role in survival. When animals and humans play games such as chasing, wrestling, or tug-of-war, they gain or maintain physical strength and learn important survival skills or proper social behavior. This is probably why play is more prevalent in younger animals: they are growing and have more to learn. But although it is important, play is not indispensable to survival. It occurs only when needs such as food, shelter, and safety are sufficiently met and when unpleasant feelings such as pain, fear, and anxiety are minimal or absent. A variety of wild and captive studies confirm that animal play is suppressed in times of hardship, such as those marked by food shortages, harsh climatic conditions, social upheaval, and stress.[1] Thus, play is a good indicator of an animal's well-being.

For many of us, the most familiar animal play is that of domesticated dogs and cats. My two cats play-stalk, -chase, and -wrestle each other. They also enjoy sporting with laser pointers, Cat Dancers, and pull toys, especially when I drag the toys beneath a rug. As unique individuals, they show idiosyncrasies in their play preferences. Megan enjoys chasing a ball of aluminum foil (preferably after I've thrown it), batting it with her front paws. If I spread newspapers out on the floor, it usually isn't long before Megan is crawling under them, and by day's end they are often crumpled and torn. Mica favors hanging upside down from the edge of an armchair, holding himself in place with his hind feet braced against an armrest, while I drag and toss a string toy within his reach.

Previous spread: **AFRICAN ELEPHANT** (*Loxodonta africana*), Masai Mara National Reserve, Kenya. Two five-year-old male African elephants take a pause from foraging to play an affectionate game of push-and-pull. Developing strength through irresistible play will serve them well later on, when they may compete physically with other bulls for access to female elephants. Photo: Yva Momatiuk + John Eastcott/Minden Pictures.

BELUGA (*Delphinapterus leucas*), Shimane Aquarium, Japan. Belugas, like their dolphin cousins, are skilled bubble-ring blowers. The rings travel through the water in a graceful vortex, and the animals manipulate them with their beaks or fins, causing them to wobble or to break into smaller rings. The behavior is primarily known in captive individuals, and it may be a way to relieve boredom. Photo: Hiroya Minakuchi/Minden Pictures.

Play isn't just for animals who eat from a bowl. Most mammals and many birds (to date, from thirteen of the twenty-seven bird orders) are known to play, and there are anecdotal accounts of playlike behavior in some reptiles and fishes. Play has been recorded in every species of primate studied.[2] On a recent visit to South Africa, I joined a "baboon walk" with the managers of Baboon Matters, a nonprofit group set up to protect the wild baboons and resolve conflicts with human residents in the Cape Town area. The rambunctious play of a group of youngsters on a grassy hillside at the end of a residential street was a joyous sight. They would scamper up the ten-meter-high slope, then leap, roll, tussle, or somersault their way down again. Flying ambushes, play bites, and limb-tuggings were part of a stream of play that went on for at least ten minutes. I could hear the thuds of their little bodies hitting the ground as they hurtled down the hill, twisting and rolling in a grappling heap of as many as four individuals. You can watch one of the sequences I filmed in a video titled "romping baboons" on YouTube; the resolution is poor, but you'll get the idea.

In the early 1980s a team of scientists in Israel spent 950 hours observing eleven groups of Arabian babblers. Babblers are songbirds similar in shape and size to the American robin or the European blackbird. They live in tightly knit family groups of half a dozen or so individuals. The researchers documented 2,500 instances of play among their subjects, including four hours recorded on videotape for analysis.[3] The most common forms of play observed were wrestling, displacement (king of the hill), chases, and tug-of-war. I've seen a segment of the tape, and let me tell you: small birds wrestling on the ground like kittens make for an arresting sight. The babblers also use several types of play signals, including first establishing eye contact and then crouching, rolling over, grabbing a stick and holding it aloft, and making a play bow. I was unable to obtain a good-quality photograph of Arabian babblers playing for this book, but you can see jungle babblers allofeeding on pages 54 and 55 and allopreening on page 80.

Because life in the wild is so often ungenerously viewed as an endless struggle for survival, we may not think that undomesticated creatures play. A correspondent to *New Scientist* magazine described seeing two magpies repeatedly take turns pecking the tip of a fox's tail before hopping off. The fox merely flicked her tail each time, perhaps out of irritation. There are many other accounts of magpies, ravens, and crows doing this sort of thing (see page 31). It's usually put in an earnest, adaptive context, and indeed another reader suggested that the magpies in her garden harassed her two cats to prevent them from approaching the tree where the birds nest.[4] But I find this explanation unsatisfying. For one thing, it isn't clear that attracting the cats' attention like this would deter them from going after the nest. For another, the second reader also reported seeing the behavior in the autumn and winter, when the nest would be empty. Such behavior could be just plain fun or exciting for the harassers—a mischievous game made more thrilling by the element of risk involved. Let me say once again that these evolutionary and experiential sorts of explanations are not mutually exclusive. One, in fact, is the genesis of the other.

Here's another fox observation that's much harder to interpret as anything but play. Sheila McGregor, a marketing consultant and dog rescuer, was bicycling on a rural road in Williston, Vermont, and noticed a Holstein cow in a pasture acting bizarre. Here's what Sheila describes seeing:

The cow was leaping in the air and cavorting around like a puppy playing. I pulled my bike over to watch. When I looked closer, I saw that she was playing with a fox! The fox was dancing around in front of her, play-bowing and then bounding away. The cow would give a playful chase and then stop, waiting for the fox to initiate again. I watched for several minutes before both animals went back to "normal" behavior.

Foxes also engage in solitary play. I recently watched one trotting across a field rimed with frost. She flopped onto her side and propelled herself along the slippery grass a few feet before getting up again. Not surprisingly, I've included an image of playing foxes (kits, in this case) in this book (see page 37).

Rich Hoyle operates the Pig Preserve, a Tennessee sanctuary for pigs rescued from various abusive and neglectful situations. He described to me some of the games these animals enjoy:

> They play tug-of-war and keep-away with large branches and small saplings, often teaming up to defeat another. They will sneak up on a buddy who is snoozing in a mud hole and splash water on him and then run like hell. . . . On a warm spring day the younger ones will run across the pasture, jump and spin around in midair just for the pure joy of it.

Some games—such as play-fighting among predators—could cause injury. This may be one reason why animals use specific signals to invite others to play. The dog's familiar play bow, for instance, is a canid's way of communicating something to the effect of: "Okay, while I might be about to clamp my jaws down on your foreleg, it'll just be playful, and I'll restrain myself from biting too hard." The play bow is recognized by animals other than dogs. There are several accounts of domesticated huskies playing with wild polar bears. One celebrated case lasted several days and was featured in *National Geographic* magazine.[5]

Different species use a variety of play signals, including a head wobble, a stiff-legged jump, a roll on the ground, a tail whip, a hunched back, and a kicking leap in the air.[6] Not all play invitations are visual; the ultrasonic chirps of mirthful rats are believed to signal readiness for positive social engagement.[7] They are an acoustic version of a dog's play bow.

Animals play fair. Detailed analyses of wrestling rats and boxing wallabies show that larger, stronger individuals actually temper the boisterousness of their actions.[8] This strategy not only averts injury (useful for the husky who plays with the bear!) but sustains the game. Animals who play too roughly may have trouble finding a playmate and risk being shunned by others. Chimpanzees who roughhouse in pairs have been shown to increase their play signaling in proximity to adults, particularly as the intensity of play bouts increases. This is thought to minimize the possibility that the mother of the younger chimp will intervene and spoil the fun.[9] All told, the evolution of play involves sophisticated communication, self-control, and fairness.

Some animals show their desire to be treated fairly by demonstrating what scientists have termed *inequity aversion*. A pair of captive brown capuchin monkeys are content to exchange a token for a slice of cucumber, but if one of them begins to get rewarded with preferred grapes, the other refuses to accept cucumbers.[10] Another recent study from the University of Vienna demonstrates a sense of fairness in dogs. Dogs who are not given a treat when they respond to a simple command (shaking a paw with the experimenter) will eventually refuse to respond to the command if a dog next to them is rewarded for the same behavior. If, however, the denied dog is unaccompanied, she or he will perform the task for much longer, indicating that the earlier refusal was not merely due to frustration or fatigue.[11]

From fairness it's not a huge leap to justice. Play has rules of conduct, encourages restraint, involves give-and-take, and fosters awareness of another's point of view. For all these reasons, some scientists now think that the roots of morality lie in animal play. Although sharing, for example, was until quite recently thought to be a solely human trait, more and more examples of animals showing fairness and consideration for others are coming to light. Dolphins, elephants, rats, and bats will come to the aid of another who is distressed, wounded, or disabled.[12] When caged capuchin monkeys have to cooperate in dragging a heavy tray so they can get the food on it, they quickly figure out how to do so, then share the effort and the food. If food is placed on one side of the tray so that only one monkey has access to it, they still share, even though that monkey could have it all to him- or herself.[13] Naked mole rats readily share their food with colonymates, and jackdaws are more likely to share favorite foods than less-preferred foods with other jackdaws.[14]

For all of its carefree nature, play turns out to have significant implications for being nice and for doing what's right. This shouldn't be too surprising. Living in social groups requires getting along, and play is one of nature's most effective social lubricants.

GRAY LANGUR (*Semnopithecus entellus*), Bandhavgarh National Park, Madhya Pradesh, India. This naughty young langur is hanging from an adult's tail and had been swinging out with it. At one point she (or he) tried to climb up the tail. The image says as much about an adult's tolerance as it does a juvenile's mischievousness. Photo: Elliott Neep.

POLAR BEAR (*Thalarctos maritimus*), Churchill, Manitoba, Canada. Two adult polar bears engage in a bout of wrestling. During autumn before the sea ice freezes, the bears spend much time sparring and play-fighting, which helps them develop predatory strength and skill. It also stabilizes the young bears' relationship until they become independent adults. But few if any biologists would contend that the bears are pondering these weighty benefits. Like us, they play because it is fun. Photo: Charleen Baugh.

HOODED CROW (*Corvus corone cornix*) and **HERRING GULL** (*Larus argentatus*), Germany. Two hooded crows look on as a third tweaks the wingtip of a herring gull. Crows, ravens, and magpies are well known for teasing other birds and mammals by pulling at a convenient extremity. These efforts may displace the target animal from food but other times appear simply to be a game. Photo: Konrad Wothe / Minden Pictures.

GRAY-HEADED FLYING FOX (*Pteropus poliocephalus*), Bellingen Island, New South Wales, Australia. Like most mammals, baby flying foxes are highly dependent on their mothers. They begin to fly at about four months and are weaned at six months. Individuals recognize one another, and they may form close friendships during their long lives (up to twenty-three years or more). Wrestling and play-fighting are common components of their play. Here a mother and her pup engage in a bout of play-fighting. Photo: Vivien Jones.

HOARY MARMOT (*Marmota caligata*), Mount Rainier National Park, Washington. A family of hoary marmots takes time out for some rough-and-tumble play at Mount Rainier National Park. The photographer observed many chases during the several hours that she observed them. Periodically, one of the two adults (only one can be seen here) would climb onto a rock to scan for predators before resuming the play. Contrary to their undeserved reputation as "lower" mammals, rodents are intelligent, emotional, and perceptive. Photo: Nancy J. Wagner.

DOMESTIC CAT (*Felis catus*), Buchs, St. Gallen, Switzerland. A British shorthair kitten demonstrates one of the oldest forms of self-amusement: object play. Photo: Dennis Lorenz.

ASIATIC LION (*Panthera leo persica*), Gir National Forest Park, Gujarat, India. Once ranging from the Mediterranean to northeastern India, Asiatic lions have been reduced to a population of about four hundred by persecution, habitat loss, and water pollution. Today they are confined to the Gir Forest in northwestern India. Fortunately, these playful youngsters are oblivious to these grim facts. Photo: Steve Mandel.

RED SQUIRREL (*Tamiasciurus hudsonicus*), Tete Jaune Cache, British Columbia, Canada. Squirrels can often be seen romping, frolicking, and chasing one another in their woodland habitats. Some chases are aggressive, but many are just playful. This pair was tearing up and down the trunk with enough speed to dislodge some bark chips (seen to the right). Sometimes one chased the other, then the roles reversed. The photographer heard the noise of their chases and watched them doing this for five minutes. Photo: Arthur Sevestre.

ALPINE IBEX (*Capra ibex*), Italian Alps. Three young male ibexes spar with their horns. Males form bachelor herds during the summer months, when they forge playful alliances to practice the maneuvers that will serve them in the more serious business of trying to establish harems in the autumn. Photo: Laura Corvini.

RED FOX (*Vulpes vulpes*), Clear Creek, Colorado. Fox kits spend a large proportion of their time playing, usually with the highest-caliber toys available: one another. Through activities such as chasing, fleeing, pouncing, biting, falling over, and wrestling, they learn vital skills for social and predatory behavior as adults. That is the adaptive basis for play. The immediate reason they do it is that they enjoy it. Photo: Allen Thornton.

BELDING'S GROUND SQUIRREL (*Urocitellus beldingi*), Malheur National Wildlife Refuge, Oregon. Belding's ground squirrels live in high-elevation meadow habitats in the western United States. Each litter numbers between five and eight pups. Youngsters are vulnerable to many predators, and they learn to respond appropriately to adult alarm calls within about a week of exiting their natal burrows, where they spend lots of time playing among themselves. Photo: Nate Chappell.

BONOBO (*Pan paniscus*), San Diego Zoo, California.
Bonobo mothers are devoted caregivers who
invent a variety of games for their infants. Photo:
Ken Bohn / San Diego Zoo / Minden Pictures.

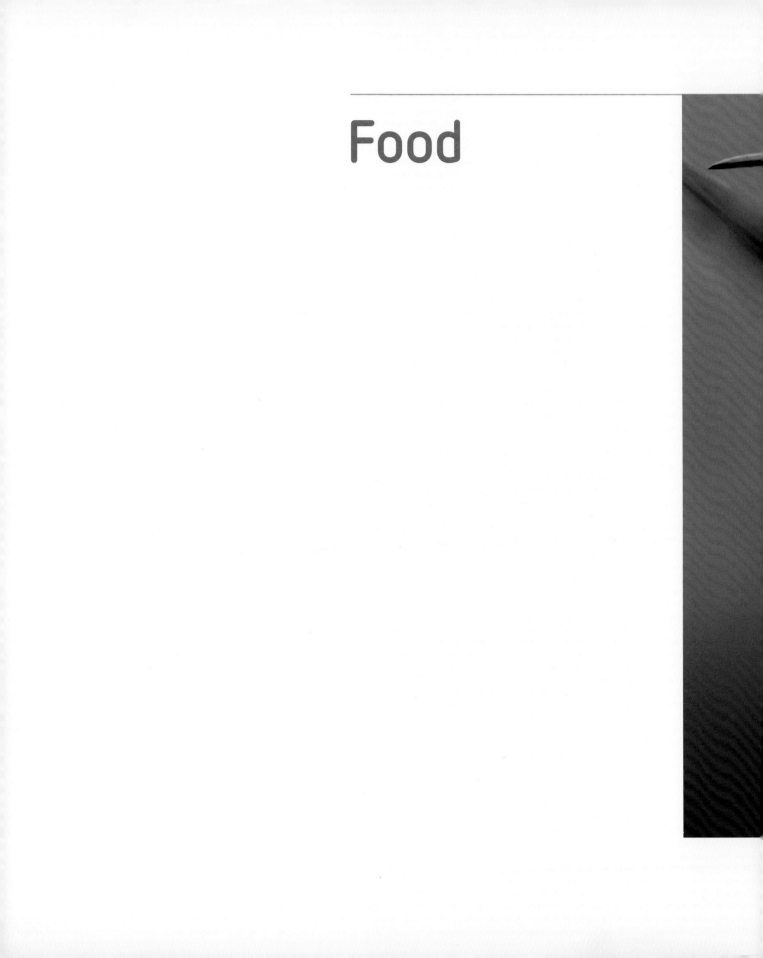

Food

Feeding oneself is one of life's basic assignments. Living requires energy, so bellies need filling. It takes a lot of nectar to fuel a hummingbird, a mother tern needs many fish meals to grow her eggs, and a busy mouse may make two hundred or more foraging trips in a single night. But securing nourishment is not all work and no play. In fact, the fundamental importance of food is a strong driver of food-associated pleasure systems. We know from personal experience that the pleasure of food includes anticipation, consumption, and (provided we haven't overstuffed ourselves) the feeling of satiation. Animals' brains release pleasure-associated opioid compounds while they search for food and while they eat it. Humans may be unique in developing recipe books, casseroles, and indigestion tablets, but we aren't alone in enjoying our food.

Animals have reactions to pleasant tastes that share evolutionary origins with and are similar in physical expression to humans'.[1] Studies have found marked similarities in facial expressions and gestures of human and animal infants in response to different tastes. Sweet tastes elicit repeated, rhythmic tongue protrusions in all species of monkeys and apes that have been studied. Rats respond to sweet stimuli with licking movements, paws drawn to the mouth, and head nodding, and to bitter tastes with a triangle-shaped mouth, lowered chin, and treadling forelimbs.[2]

In a world of food choices—some of which may be unwise—a discerning palate is useful. Spider monkeys show clear preferences for higher-energy foods. They also have high taste sensitivity, as determined from preferences shown for very low concentrations of a polysaccharide dissolved in water.[3] Four species of monkey differed markedly in their acceptance of different concentrations of

Previous spread: VIOLET SABREWING (*Campylopterus hemileucurus*), Costa Rica. Hummingbirds not only have keen senses for finding flowers but can also remember which ones they've visited and know how long it takes different kinds to replenish their nectar supply. Photo: Tim Fitzharris / Minden Pictures.

COMMON PORCUPINE (*Erethizon dorsatum*), Kalispell, Montana. Members of the vast rodent order, North American porcupines feed mainly on bark, leaves, and conifer needles, but they also include in their diets a variety of roots, stems, berries, fruits, seeds, nuts, grasses, and flowers. Like most animals, individual porcupines have particular food preferences. This one is enjoying the leaves and berries of a wild rose. Photo: Lynda Fowler.

sour-tasting water. Olive baboons preferred most of the sweet-sour taste mixtures offered them, and spider monkeys and pigtailed macaques showed intermediate preferences, while squirrel monkeys showed the lowest tolerance for sour tastes, rejecting most of the mixtures. All four species perceived both the sweetness and the sourness of the mixtures, and each had different degrees of hedonic (pleasurable) response to the two taste properties, probably reflecting differences in their natural diets.[4] These species also make fine discriminations about the presence of MSG and table salt dissolved in water.[5] Squirrel monkeys given two-bottle preference tests consistently prefer food-associated sugars (sucrose, glucose, fructose, and lactose) over tap water, and their detection threshold for sucrose is identical to that which has been established in humans.[6]

It would be a stretch to think that these sorts of preferences are based solely on energetics and without any regard for the experience of the pleasure of, say, sweet flavors. In the words of the evolutionary biologist Marlene Zuk, "it seems reasonable to assume that the proximate reward of food shapes the ultimate goal of survival."[7] Our own experiences illustrate this: for evolutionary reasons, we eat to sustain ourselves, but consciously we eat because food is enjoyable. A fruit bat who eats a mango is promoting her survival, but it is the sweet smell and taste of the fruit that she experiences (see page 53).

Studies have shown that animals behave flexibly and hedonically in the presence of food choices. Rats will enter a deadly cold room and navigate a maze to retrieve highly palatable food (e.g., shortbread, pâté, or Coca-Cola). If they happen to find their regular (and less palatable) commercial rat chow at the end, they quickly return to their cozy nests, where they stay for the remainder of the experiment. If, however, they find a tasty treat, they feed on it before returning home, then return repeatedly for more.[8] Reptiles also appear capable of making food choices based on the anticipation and experience of pleasure. Given a choice between their routinely available (and probably not very exciting) reptile chow right beneath their sun-lit perch and a gourmet tidbit—fresh lettuce—at the other end of the room, iguanas shunned the chow and ventured out to get the lettuce, even when it required them to enter a dangerously cold environment.[9] Furthermore, the lizards' willingness to retrieve the tidbit varied with the cost/benefit of making the trip; as the temperature near the lettuce was set lower, the iguanas were more likely to stay put and eat the reptile chow.

Because they are so visible to us, birds offer many opportunities to observe feeding behavior in wild animals. My home in suburban Maryland backs onto a mature woodland, and my neighbor has deployed a generous array of feeders, so I can often watch birds eat. On one occasion, a blue jay begged noisily from a branch as another hammered at a nut between his claws, bolting back hunks with a forward jerk of his head. He gave the last piece to the obsequious loiterer, who gargled out a cry on receiving it. Perhaps this was a parent-offspring tandem, or perhaps they were buddies. Another time, I watched a recently fledged hairy woodpecker calling expectantly as she watched an adult male—presumably her father—drilling at my deck near where carpenter bees had made one of their nests in the strut beneath the handrail. The clever male reached into the hole he'd made, plucked a fat grub from its chamber, and passed it off to the fledgling before finding one for himself. Woodpeckers are skilled at detecting and removing invertebrates from their woody niches, which offers them a year-round supply of edibles—especially important during the cold winter months, when most insects are sequestered away (see page 59).

Different species often have idiosyncratic approaches to food. I always thought starlings were just grasping at things when they poked their pointy beaks into the grass, until I noticed that they actually move the blades away from one another by opening their beaks, so they can look below. At the botanical gardens in Sydney, Australia, I watched a small flock of sulphur-crested cockatoos foraging along a lawn. They pushed their beaks into the grass and selectively removed individual grass seedlings. I could hear the faint snapping sounds as roots were torn up. As best as I could tell through my binoculars at close range, the birds were nibbling only at the very base of each sprig before dropping it and selecting another. I retrieved two fresh discards; one happened to be intact, while the other had lost about a centimeter of stem from the base. I tasted the intact one; it was succulent but flavorless to me. Yet it would be naïve to conclude that a food item whose taste we find unrewarding is similarly unpalatable to another species. Animals' sensory systems are adapted to their niches, and that includes the types of food they eat.

Like us, animals tend to be particular about what they will and will not eat. At the far end of the same field on which I'd watched the cockatoos, a rainbow lorikeet landed just a few feet away from me. I discovered a package of dried fruit in my shoulder bag. I bit off a small chunk of apricot, kneeled, and offered it to the lorikeet, who calmly walked up, gently grasped the morsel with her beak, and touched it with her tongue. She rejected the fruit, releasing her grip and decidedly walking away. I offered the apricot again, but she didn't return. However, when I held out a small piece of prune, the lorikeet came straight over and tasted it. This tidbit also failed her taste test, but apparently she recognized from a distance that the dark morsel was different from the orange one and therefore worth a try. The interaction illustrates not only keen awareness on the bird's part but a discerning palate too. I learned later that wild rainbow lorikeets feed mainly on pollen and nectar, which they collect with a spongy tip on the end of the tongue.

Food preferences vary among individuals as well as species. Pallid bats studied in California, for example, showed individual differences in the types of insects they ate, though the study authors do not address the possibility that these patterns may be based on palatability preferences, suggesting instead that prey availability and individual foraging behavior are responsible.[10] In a study of captive parrots at the Jersey Wildlife Preservation Trust in the United Kingdom, wild-caught birds liked more types of food than did captive-born birds.[11] If you've cohabited with two or more dogs or cats, you've probably noticed idiosyncrasies in their dining preferences. My cat Megan is more eager than Mica to lick Marmite (a salty yeast-extract spread high in B vitamins) from my finger, but if I substitute Vegenaise (a vegan mayonnaise), Mica is first to arrive and last to depart. I kept three rats while living in England a few years ago, and they also revealed individual preferences. Rats are omnivores with catholic tastes. In addition to showing interest in anything new that I offered them, Lucy, Rachael, and Veronica all loved peanut butter cookies, two of them enjoyed frozen peas, and only one was fond of hummus.

Individual preferences also change through time. I disliked tomatoes and avoided brussels sprouts as a boy, but now they are among my favorite foods. Olfaction, which involves smell and taste, is known to be the most malleable of the senses.[12] With just a few exposures we can become completely accustomed to new foods. Many years ago I decided to remove animal products from my diet, and I remember noticing at first the different flavor of soymilk compared to the cow's milk I was used to. Within a few days, however, the soy alternative seemed completely normal. This flexibility in taste preferences may relate to maintaining nutritional balance: we are more likely to meet our nutritional needs with a more varied dietary base, so it may be adaptive to have a malleable palate. It may also be useful for replacing food sources that fluctuate seasonally. Some trees only bear fruit or nuts for a few weeks every two or three years. Frugivorous birds and mammals may gorge on them during these bonanzas, then switch to different fare as the trees stop producing.

In some fish species, like the brown bullhead of the catfish family, the entire body surface is cloaked with taste sensors. It's a useful adaptation when you spend much of your time in murky waters where vision isn't very effective. Carp, suckers, and the aptly named barbel (to name a few) also have highly taste-sensitive whiskerlike barbels around their mouths, which help them discern palatable from inedible materials as they vacuum along the mucky bottom. Studies on minnows conducted in the 1920s found that they could be trained to discriminate among not only sweet, salty, sour, and bitter substances but also several natural sugars.[13] Recent research has shown that fishes share with mammals components of taste perception and that the two vertebrate groups distinguish desirable from aversive tastes along common sensory pathways.[14]

Finally, the pleasure of food can have a social dimension. Many birds offer tidbits to prospective mates, and some, like the tern on pages 60–61, may feed a partner throughout the nesting period. Food plays an important role in the social dynamics of primate societies. Wild chimpanzees and bonobos share prized food items to secure friendships and boost status.[15] There's no denying the indispensability of food to individual survival. When gustatory pleasure is added, little wonder that food is so energetically sought after and valued by humans and animals alike. George Bernard Shaw, in *Man and Superman*, said: "There is no love sincerer than the love of food." Seeking it out gives purpose and meaning to the lives of countless animals. Eating it is the most life sustaining of activities. And for the sentient forager, food can be a source of pleasure.

ARCHERFISH (*Toxotes jaculatrix*), Singapore. Famed for the accuracy with which they can knock insects from overhanging vegetation by spitting water from their mouths, archerfishes like this one sometimes opt to leap for their prey. Contrary to popular myth, fishes are fast learners and remember things for months or years. They also have a good sense of taste. It follows that they may anticipate and enjoy a good meal. Photo: Scott Linstead.

HOARY MARMOT *(Marmota caligata)*, Mount Rainier National Park, Washington. This hoary marmot wandered through a patch of plants, sniffing at each flower before eating it. From a survival perspective, one might say that the marmot was making sure the flowers were ripe and edible. From an experiential perspective, he was probably savoring the aroma of his next meal. Photo: Nate Chappell.

GRAY LANGUR (*Semnopithecus entellus*), Sasan Gir, Gujarat, India. A mother langur and her child enjoy some digestive biscuits tossed to them at a railroad station in western India. Photo: Steve Mandel.

EGYPTIAN FRUIT BAT (*Rousettus aegyptiacus*), Kruger National Park, South Africa. Despite their superficial similarity, the fruit bats (suborder megachiroptera) differ profoundly from the insectivorous bats (suborder microchiroptera). Lacking echolocation, this individual relies on keen night vision and smell to locate food like this ripe, succulent mango. Photo: Merlin Tuttle / Bat Conservation.

JUNGLE BABBLER (*Turdoides striatus*), Gir National Forest Park, Gujarat, India. Babbler parents continue to feed their babies after they have fledged and left the nest. This youngster assumes a characteristic begging posture, with arched back and fluttering wings, while an adult stuffs a tidbit into its gaping mouth. The piercing white iris of this species lends a serious intensity to their demeanor, but it can be deceiving: babblers are among the most playful of songbirds. Photos: Steve Mandel.

DOMESTIC PIG (*Sus scrofa domestica*), Farm Sanctuary, Watkins Glen, New York. These piglets were born to one of several factory-farmed sows rescued from the 2008 Iowa floods by Farm Sanctuary, an organization that cares for neglected and abused farmed animals and conducts public education campaigns. Prior to her rescue, the sow, stranded on a levee, made a nest out of whatever she could find. Photo: Siobhan McClory.

AMERICAN PIKA (*Ochotona princeps*), Rocky Mountain National Park, Colorado. Pikas live among mountain rocks, where they store food for the colder months. They sometimes leave gathered vegetation in the sun to dry, presumably to increase its shelf life. In Eurasia, pikas often share their burrows with snowfinches, who build their nests there. This individual looks all business but may be feeling the sense of satisfaction that comes from getting an important task done. Many studies show that rodents and other animals prefer foraging to eating from a food dish and that they can develop psychological illnesses when chronically deprived of opportunities to perform important survival behaviors. Photo: Bill Meikle.

CEDAR WAXWING (*Bombycilla cedrorum*), East Lansing, Michigan. One might look at this cedar waxwing and think that birds fill their stomachs only to stave off hunger. A less impoverished view holds that they are drawn by the bright colors and enjoy their meal of berries. Birds are particular about what sorts of food they prefer. This one is gorging on crab apples, which also happen to be more nutritious than the commercial apples we eat. Photo: Dennis Lorenz.

BUTTERFLY (Polyommatinae sp.), Balkan mountains, Bulgaria. A lump of ferret dung is a magnet to these butterflies, known as "blues," who are drawn to the concentration of salts and other nutrients. Photo: Andrey Antov.

GREAT SPOTTED WOODPECKER (*Dendrocopos major*), northern Denmark. This female great spotted woodpecker listened for a while before excavating the beetle larva seen in her beak. I have watched a similar-size hairy woodpecker chiseling into my wooden deck before plucking a fat carpenter bee larva from the hole and passing it to a fully fledged chick who was begging insistently nearby. It is easy to imagine that these alert birds start to anticipate the impending treat when they hear sounds emanating from the wood. Photo: Anders Grøndahl Nielsen.

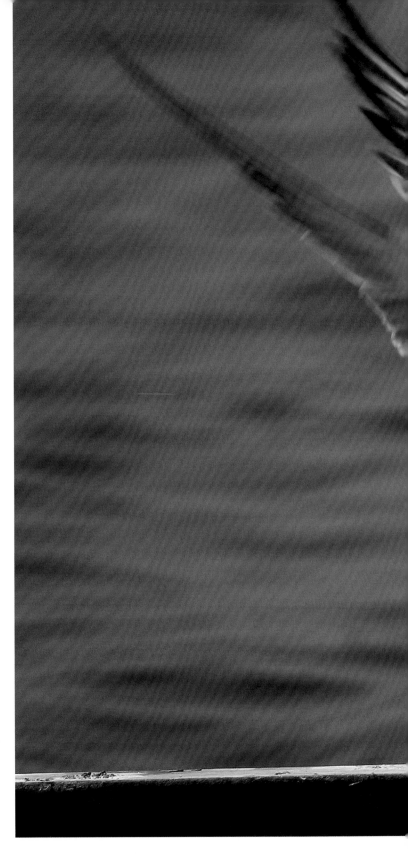

COMMON TERN (*Sterna hirundo*), Gan Shmuel, Israel. Terns often practice what is termed courtship feeding, in which one bird—typically the male—plies his mate with fish. It is an example of how evolution often leads to virtuous behavior: males show their commitment to the females who will bear their young by investing in gifts. Females, weighed down by developing eggs, benefit by getting food without having to work hard for it. Photo: Danny Laredo.

Touch

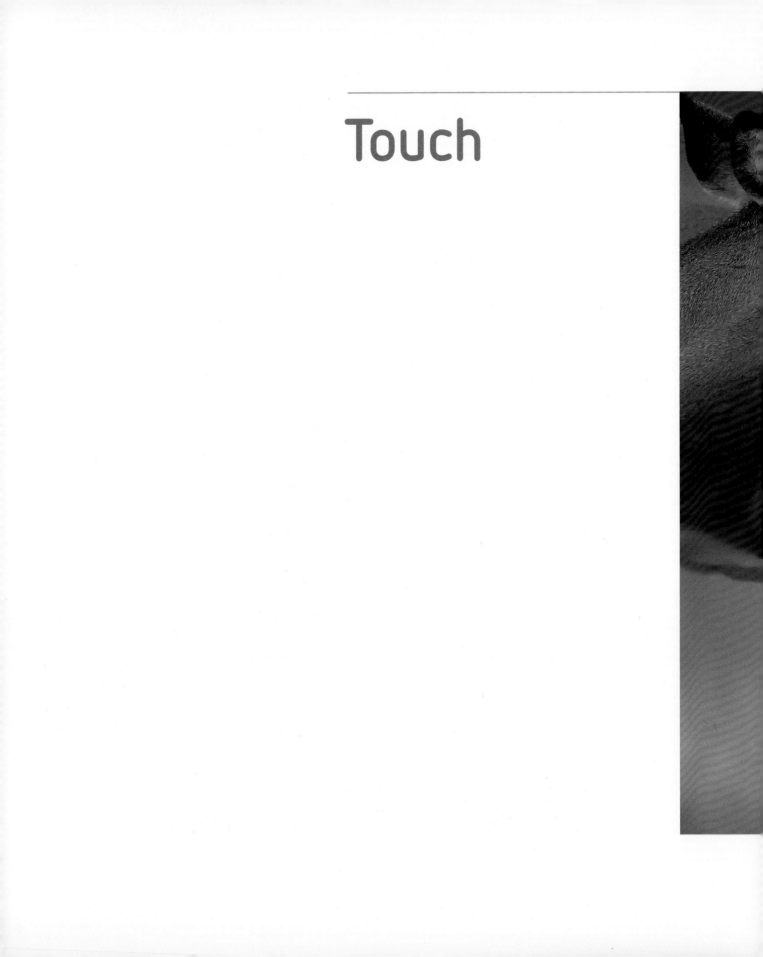

My home in the suburbs of Washington, DC, is situated about seventeen miles from a sanctuary for formerly abused or neglected farmed animals. Poplar Spring Animal Sanctuary occupies four hundred acres of beautiful rural countryside abutting the Potomac River. Walnut and persimmon trees dot the winding, mile-long gravel driveway. A pair of bald eagles has nested for twenty years high atop an oak tree near the original homestead, which dates to 1760. In its ten years of operation, Poplar Spring has flourished in the capable hands of its founders and full-time residents, Terry Cummings and Dave Hoerauf. Among its four hundred domesticated denizens are pigs, cattle, sheep, goats, chickens, turkeys, guinea fowl, rabbits, horses, and a pair of thirty-three-year-old mules. All have names. And all are meticulously cared for by the handful of part-time paid staff and a cadre of dedicated volunteers.

One sunny but chilly winter morning, my daughter and I stood among a throng of sheep and goats, having just finished sweeping out their heated stalls. Two of the tamer sheep—Clover and Hickory—stood between us. At that time of year their fleeces are full, and it was the most natural thing for Emily and me to plunge our cold fingers into the spongy mats in front of us. The sheep's bodies felt warm beneath two inches of dense insulating wool. I began massaging my fingers along Clover's back. It felt nice, and Clover didn't seem to mind. After a couple of minutes, I felt something scraping against my boot. I looked down and saw it was Clover's front hoof. I realized I had stopped massaging her. My interpretation is that Clover enjoyed the massage and was letting me know she wanted me to continue.

Why was Clover enjoying the massage? It's not as though sheep go around giving one another back rubs. But like ours, their muscles work hard, and they probably get tired and sore. So I'm guess-

---

Previous spread: **WHITE-TAILED DEER** (*Odocoileus virginianus*), Cook County, Illinois. Deer are very tactile animals and can often be seen licking one another. It feels good and is a way of expressing affection. Photo: Vic Harris.

GRIFFON VULTURE (*Gyps fulvus*), Artis Zoo, Amsterdam. We don't usually associate vultures with feelings like affection. But like any long-lived animal with a social life, they form attachments and show consideration for others. Here, one individual grooms another, perhaps his or her mate. Photo: Merel Den Daas.

ing it feels good to have them stimulated this way. Cats and dogs also don't give one another massages, but anyone who's lived with them knows they usually enjoy being stroked and rubbed.

As individuals started forming societies, their tactile senses took on new roles. Touch became more than just a means of perceiving one's physical environment (temperature, wind, solid objects, etc.); it became a tool of communication. Touch can defuse a conflict, comfort a victim, or say *I trust you*. Pleasurable touch usually occurs in a social context. Most of the images in this chapter involve two or more animals engaging in one-way or mutual touch.

Animals—including humans—also use touch to express camaraderie, fondness, and love. Jennifer Margulis has this to say of the giraffes she has been observing and studying in Niger for fifteen years: "They are so affectionate . . . weaving their necks in and out and rubbing up against each other—just constantly physical and touching each other. It's almost like they're doing some kind of intricate ballet. To see the affection they have for each other—it's just so beautiful."[1]

Grooming one another (*allogrooming* in scientific jargon) is an important activity for many mammal species. Some primates spend 20 percent or more of their waking time grooming or being groomed.[2] Monkeys and apes usually groom with their dexterous hands, combing efficiently through another's fur in search of parasites or bits of salt, which they sometimes eat. Hoofed mammals such as cattle, horses, and deer use their tongues, as do dogs, cats, hyenas, and other members of the carnivore family. Many rodents, including rats and mice, also spend a lot of time allogrooming, and they may use either their tongues or their paws.

In birds, preening replaces grooming. And once again, allopreening is commonly engaged in by species whose members spend long periods in one another's company. Parrots and penguins, for example, are long-lived and must invest heavily of their time and energies to successfully rear their chicks. On average, more-proficient parents rear more chicks to adulthood. It is now widely accepted that the pleasure of each other's touch helps to cement and sustain the emotional bond between mates, and closer pairs tend to work more cooperatively. Thus, natural selection favors affectionate behavior in these populations.

One way to measure pleasurable responses is to monitor an animal's heart rate. In a study of partially tame Camargue horses, external heart monitors were taped to their bodies and the animals' heart rates scored during bouts of brushing by trusted human researchers. A slowing of the pulse indicates a relaxed state and, probably, feelings of pleasure in these horses, which often lick and nibble one another in particular areas of the neck and withers (see page 79). The hypothesis was that the animals derive more pleasure from being allogroomed in these specific areas. True to prediction, the horses' pulses slowed significantly more when the researchers focused their brushing on these favored areas than when the brush was directed at other regions of their neck and withers.[3]

The pleasant sensations of touching encourage individuals to do it again. Such actions tend to be evolutionarily beneficial because rewards—like pleasure—are a powerful way to inspire and reinforce adaptive, or "good," behaviors. For example, baby iguanas often rest in groups with their heads and long tails draped over one another. Sunbathing marine iguanas also often rest against each other (see page 71). There are probably survival benefits to this, such as enhanced predator detection and temperature regulation. But, as the iguana expert Gordon Burghardt asserts, such adaptations needn't exclude feelings of security that can be pleasurable and relaxing.[4] I would add that the two go hand in paw—that the good feelings are a product of their adaptiveness.

In his poem "Fish," D. H. Lawrence imagined the tactile world of the titular creature as *a sluice of sensation along your sides*. It is difficult to know how fishes might be feeling, but their behavior—like that of the sheep at Poplar Spring Animal Sanctuary—can provide clues. One of the best-studied and most intriguing mutualisms in aquatic environments is that of cleaner fish and their clients. These relationships occur in marine and freshwater habitats, but the best-known one happens on reefs, where small, brightly colored wrasses perform a cleaning service for larger fishes, who line up to wait their turn. It's rather like a visit to a barber or a manicurist: we tend to return to a favorite practitioner, we are prepared to wait for their service, we trust them, and we pay them. Reef fishes also develop loyal cleaner-client relationships. Cleaners signal that they are open for business by swimming vertically, and client fishes queue for their turn with their preferred cleaner and cooperate by opening their mouths and their gill covers (see the interaction between cleaner wrasses and different clients on pages 15 and 75). There is mutual trust—cleaners will not bite or be bitten by a customer—and cleaners are paid in the form of sloughing skin, parasites, algae, or other contaminants that they remove from the client's body and eat. These fishes are very busy: a single cleaner wrasse can have more than 2,300 interactions per day with clients of various species.[5]

Why do fishes line up to be picked over by cleaner fishes? If you think it is because they know it is good for them to have parasites and algae plucked from their scales and gills, then you are probably ascribing more awareness to a fish than they are commonly given credit for. But there is mounting evidence that fishes are intelligent and aware, plus they receive a more immediate benefit: the tactile ministrations of the cleaner fishes feel good.

I've read several papers on cleaner-client fish mutualisms but have yet to see any specific mention of the word *pleasure* in helping to sustain these communities. The closest the authors come to acknowledging the proximate experiences of the fishes is to show that "tactile stimulation" has benefits. For example, piscivorous (fish-eating) client fish become less aggressive—and hence less threatening to other fishes in the vicinity—when they have been cleaned. Furthermore, the rate at which piscivorous clients chased prey was inversely proportional to the amount of tactile stimulation the predator had received from the cleaner.[6] This result suggests that the cleanings have a calming effect on client fishes, an outcome compatible with the hypothesis that the cleanings are pleasurable for them.

Other facets of this mutualism support a role for pleasure. Clients are choosy, selecting cleaners who provide high-quality service.[7] Some fishes change color while being "serviced" by cleaners, and there are many anecdotal accounts suggesting that color changes in fishes may reflect changes in emotions (e.g., stress, arousal). Touch also appears to be an important motivator for these interactions. A study involving 112 hours of surveillance of twelve different cleaners revealed that they seem able to influence how long a client decides to stay for an inspection, and to stop clients from fleeing or retaliating for a bite that made them jolt, by applying tactile stimulation in the form of stroking with their fins.[8]

Incidentally, cheating does occur in this system. Some fishes pose as cleaners and then, when the time is right, nip a piece from a client's fin and dart away. Redouan Bshary at the University of Neuchâtel, Switzerland, has found that clients monitor individual cleaners and develop "image scores" based on the quality and reliability of their services. A particular cleaner may behave as a "normal" cleaner at her own station but as a biter when she visits another cleaner's area. This suggests that a cleaner fish's behavior can be completely independent of her internal state.[9]

The wildlife photographers Mark Deeble and Vicky Stone have documented a similar mutualism in Kenya's Mzima Springs. Following a night of grazing on land, hippos return to the cool waters, where they are mobbed by fishes of various species who come to nibble on different parts of the hippos' bodies. Narrow-snouted barbels are able to clean between the hippos' toes and beneath their tails, cichlids vacuum their bristles, and labeos act as toothbrushes, polishing the hippos' massive teeth. The hippos are active participants, spreading their toes and legs and gaping their mouths (see page 74). Some become so blissfully relaxed that they drift off to sleep.[10]

As if not to be outdone by their hippo cousins, warthogs at Uganda's Queen Elizabeth National Park flop down on their sides to be swarmed by mongooses. The little predators love to nibble at the warthogs' skin, perhaps for salt or external parasites. The warthogs show every sign of thoroughly enjoying the attention, often stretching out with their eyes closed. In parts of India, feral dogs have forged a relationship with local troops of langur monkeys, who groom the dogs. It isn't clear why these interactions occur from an adaptive standpoint, but both parties appear to enjoy it (see page 77). Cross-species interactions like these show that nature is malleable and opportunistic. Animals are not rigidly fixed in their prescribed ecological roles. Lions really can lie with lambs, and touch can be the mediator.

MARINE IGUANA (*Amblyrhynchus cristatus*), Academy Bay, Santa Cruz Island, Galápagos Islands, Ecuador. These gregarious lizards often huddle up close, whether they are clinging to rocks as the surf breaks over them or merely sunning themselves. Here a male and a female appear to be enjoying the pleasure of their beneficial contact. Photo: Tui De Roy / Minden Pictures.

SCARLET MACAW (*Ara macao*) and RED-
FRONTED MACAW (*Ara rubrogenys*),
Dominican Republic. Members of the
parrot family, like these macaws, live for
decades and form close relationships
with other individuals. Allopreening
(preening another) is one of the ways they
express affinity. Photo: Neil Bramley.

HIPPOPOTAMUS (*Hippopotamus amphibious*) and RED-EYE LABEO (*Labeo cylindricus*), Mzima Springs, Kenya. At Kenya's Mzima Springs, hippos enjoy a daily spa treatment, being picked over by fishes of various species who remove dead skin, parasites, dental plaque, and scraps of vegetation. Far from passive recipients, the hippos spread their legs, splay their toes, and hold their mouths agape to facilitate the fishes' efforts. Photo: Mark Deeble and Victoria Stone / Minden Pictures.

ORCA (*Orcinus orca*), San Juan Island, Washington. This behavior, termed *kelping*, involves dragging seaweed (especially kelp) over a body part, usually the tail and often in the grooved portion between the tail flukes. We aren't certain why orcas do this, but they are strongly tactile, and it is probably pleasurable. Photo: Monika Wieland.

MAP PUFFERFISH (*Arothron mappa*) and BLUE-STREAKED CLEANER WRASSE (*Labroides dimediatus*), Solomon Islands. The relationships between cleaner fishes and their clients are complex. They are mutually beneficial—cleaners get nourishment, and clients get a spa treatment—and the interaction is probably pleasurable, at least for the client. Photo: Chris Newbert / Minden Pictures.

JAPANESE MACAQUE (*Macaca fuscata*), Jigokudani Monkey Park, Nagano, Japan. A hot spring on a winter day is blissful enough. The only thing that could make it better is a massage. Photo: Andrew Forsyth.

NORWAY RAT (*Rattus norvegicus*), Tucson, Arizona. Highly social and tactile, rats make excellent companions. This rat, named Delphiniums Blue for his coloration, would enjoy relaxing sessions on the front porch with his human guardian. He'd either lie, as seen here, with his back up against Brandi's legs or sleep in the crook of her neck. These intimate sessions often left Delphiniums bruxing (tooth-grinding), a behavior commonly expressed by contented rats. Photo: Brandi Saxton.

GRAY LANGUR (*Semnopithecus entellus*) and **DOMESTIC DOG** (*Canis familiaris*), Rajasthan, India. One doesn't normally expect domestic dogs and monkeys to interact socially, but it happens in Rajasthan, India (among other places), where wild langurs groom feral canines. The dogs probably benefit by having some parasites removed and the monkeys by getting a few extra tidbits. It is also probably pleasurable for both parties. Photo: Cyril Ruoso / Minden Pictures.

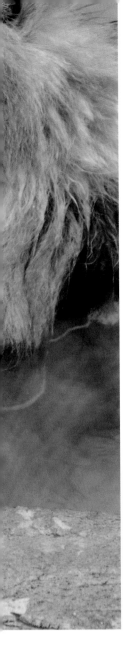

BARBARY MACAQUE (*Macaca sylvanus*), Galta Temple, Rajasthan, India. Some wild primates spend a fifth of their waking time grooming one another. They become deeply absorbed in the activity, and the recipient's bliss is almost palpable. The pleasure of touch is what makes grooming so effective as a social lubricant. Photo: Andrew Forsyth.

KONIK POLSKI HORSE (*Equus ferus*), Oosmaarland, Belgium. Horses, zebras, and other members of the horse family are known for this characteristic pose, in which a pair nibble and groom each other's neck and withers. A study using external heart rate monitors showed that the animals relaxed most when human groomers focused on the horses' favorite spots. Photo: Jonathan Lhoir.

JUNGLE BABBLER (*Turdoides striatus*), Keoladeo National Park, India. Babblers live in tight-knit social groups. These playful birds have leisure time, which may be spent preening one another. Photos: Niranjan Sant.

LITTLE OWL (*Athene noctua*), Habesor Stream, Israel. Owl parents help each other to raise their young, and they express their devotion with tender nibbles. Photo: Danny Laredo.

Courtship and Sex

Few people are acquainted with much scientific information about animal sexual behavior. And the chances are good that whatever you may have learned wasn't about pleasure. It's a recurring theme in this book that science rarely places animal behavior in a pleasurable context. And sex is no exception. As a rule, scientists treat animal sex in strictly evolutionary terms. It's not that a scientist would necessarily deny that animals may enjoy sex. Rather, it's just that the sensual—dare I say *erotic?*—nature of reproductive biology usually goes unexamined.

There are some notable exceptions. I remember well a lecture given by the prominent ethologist Frans de Waal that I attended as a graduate student. The topic was sexual behavior in bonobos—the close cousins of humans and chimpanzees that are featured on this book's cover. At the time I was not well versed in bonobo natural history. Professor de Waal included many film clips that graphically illustrated the important role of sexual activity in this species, and there could be no denying that these apes are highly sexed. Mine was not the only red face in the room that day.

For bonobos, as for humans, sex functions much more than as a mere procreative activity. Sexual pleasure is interwoven into the very fabric of a bonobo's waking existence. It is a social lubricant, practiced in various forms by all ages and gender combinations. Like touch, sex may soothe differences and defuse tensions.

Bonobos (and chimps) also use sex as a bartering tool. For example, a female bonobo, on seeing a male with two oranges, may present herself for sex and afterward walk away with one of the fruits.[1] Conversely, bonobos may barter good deeds for sex. A lower-ranking male seeking a tryst with an equally receptive female may try grooming the alpha male to earn credit for that privilege. After

---

Previous spread: **EUROPEAN BROWN FROG** (*Rana temporaria*), Vlaardingen, the Netherlands. Frogs and toads are enthusiastic about sex. Males get so carried away that they sometimes mount other males, who have a special call to alert the wayward suitor of the error he has made. Photo: Jasper Doest / Photo Natura.

SPINNER DOLPHIN (*Stenella longirostris*), Kealakekua Bay, Hawaii. Dolphins are intensely social and sexual beings. They rub up against each other, they insert body parts into others' genital slits, and they stimulate one another by directing intense pulses of ultrasound at genital areas. Photo: Robert Parnell.

the grooming session, the suitor approaches the female, and, with luck, they mate. All the while, he keeps an eye on the alpha, and if the alpha shows signs of agitation, the suitor may seek to appease him with another grooming session. This is not meant to imply that the female is a passive participant. The appeasing male isn't likely to go to the trouble unless he knows the female focus of his desires is also desirous of his company.

Female bonobos are more libidinous than males. They regularly pair up to perform what is called genito-genital rubbing, or GG rubbing. In GG rubbing, the two (occasionally three) participants face each other in the missionary position. The two clitorises—which are larger and more externalized than in most mammals—are brought into contact, then rubbed together rapidly for ten to twenty seconds. The behavior, which may be repeated in rapid succession, is usually accompanied by grinning, shrieking, and clitoral engorgement. On average, a female bonobo engages in a bout of GG rubbing about once every two hours.[2] Like humans, bonobos usually mate face-to-face, and the evolutionary biologist Marlene Zuk has suggested that the position of the clitoris in bonobos and some other primates has evolved to maximize stimulation during sexual intercourse.

There are good reasons to expect that sex is fun for sentient animals. Although sex (unlike food, for example) is not necessary for individual survival, for genetic survival it is indispensable. Most biologists speak of *reproductive success* as the ultimate measure of an individual animal's evolutionary fitness. We may expect that evolution favors animals who are highly motivated for sex. And pleasure is a powerful motivator.

Animals are equipped for sexual pleasure. Genital areas are highly innervated, and all female mammals have a clitoris, which becomes engorged in those relatively few species in which this has been studied. There is also evidence that bonobos aren't the only nonhuman primate females who experience orgasm. A "climax face" has been described in several species of monkey, accompanied by rhythmic pelvic and vaginal contractions, genital engorgement, and increased heart and breathing rates.[3] Among the adaptive bases for suggested female orgasm are that it helps to propel sperm toward an egg and to cement the bond between parents who must work together to raise young.[4]

In 2001 a research team at the University of Sheffield described the first evidence of orgasm in a bird. For fifteen or so minutes prior to mounting, the male red-billed buffalo weaver rubs a penislike appendage against a female. During this time he becomes increasingly animated and appears to reach a state of sexual climax.[5] Homosexual and autoerotic behaviors have been observed in at least 120 different bird species.[6] What we haven't yet noticed in the remainder may reflect our ignorance, not an absence.

Besides GG rubbing, there are many cases in which animals engage in sexual activity that has no chance of successful procreation. Examples include mating outside a well-defined breeding season, mating during menstruation, same-sex mating, manual or oral stimulation of another's genitals, and masturbation. The Seattle-based biologist Bruce Bagemihl has compiled scientific documentation on nonprocreative sexual behavior in mammals and birds into *Biological Exuberance: Animal Homosexuality and Natural Diversity*, a meticulously researched tome of more than seven hundred pages.[7] To leaf through this book is to encounter evidence of pleasure-generating sexual behavior on nearly every page. Photos and illustrations depict—among other things—bonobos sharing an open-mouthed kiss, a walrus stimulating his erect penis with a flipper, an orangutan masturbating with a piece of liana, a pair of manatees embracing in a 69 position with each male's penis in the other's mouth, a young red fox mounting her mother, and a white-tailed deer rubbing his penis against his rib cage.

"The Sex Lives of Animals," an exhibition shown at the Museum of Sex, in New York City, from July 2008 to March 2009, also assailed the common myth that animal sex is just about procreation. Animals engage in "every kind of penetrative intercourse imaginable," wrote one reviewer of the display,[8] for which I was pleased to serve as a scientific advisor. Gender itself is changeable in many animal species, especially fishes. Conversions from female to male are known in at least fourteen fish families, and from male to female in at least eight.[9]

The technical term for this reproductive strategy is *sequential hermaphroditism*. Some other animals are *simultaneous hermaphrodites*, meaning they assume the roles of both male and female at the same time. One of my biology mentors spent several years studying land snails in the southeastern United States. These slow-moving mollusks compensate for their sluggishness (pun intended) with the ability to mate with themselves. Fewer of their eggs hatch, but during lean times it's better to be a hermaphrodite than a lonesome failure at love. Am I suggesting that snails get a kick out of sex? Not necessarily, but you may want to reserve judgment until you've seen them copulating (see page 97). Several fish species are now also known to be simultaneous hermaphrodites, including sea bass, hamlets, gobies, and sandfishes.[10]

If sex were about nothing more than making babies, why would so many animals engage in manifestly wasteful sex acts? It might be argued that the participants are confused and don't know their efforts will be fruitless. But that interpretation assumes animals understand that sexual intercourse begets young—a presumption generally rejected by scientists.

Most of us are taught to believe that sex is merely an instinctive act for animals. Not so. Animal sex is versatile, opportunistic, and sometimes creative. It's true that males of most species have the potential to produce many more offspring than females, whose fecundity is limited by a greater energy investment in making babies (e.g., gestation, incubation, lactation, other parental care). But it doesn't follow inevitably that males eagerly vie for coy females, and it isn't just males who pursue females. Elliott Neep's image of a red deer doe mounting a stag (page 99) illustrates how animals may express sexuality outside stereotyped roles. During my walk with a wild baboon troop in South Africa I watched lustful young females exposing their bright red estrous swellings to males—clear invitations for sex.

I am really delighted to include images of courting and mating sharks (see pages 96 and 97). These behaviors are rarely seen and even more rarely photographed. Sharks rival bats, snakes, and spiders as animals unfairly burdened by negative publicity. And as with virtually all demonized animals, reality doesn't support their reputation. Human fatalities due to shark bites are exceedingly rare, averaging fewer than ten per year worldwide. (Compare that to the twenty-five million to seventy-five million sharks killed by humans yearly, and you may wonder who deserves the meaner reputation, them or us.) Mating in most (perhaps all) sharks involves courtship, during which the male communicates his amorous intentions to a sexually receptive female with ritualized swimming and gentle "love nips" along her back or flanks. With their ventral (abdominal) surfaces united, the male inserts one (rarely both) of his paired "claspers"—intromittent organs, which deliver sperm—into the female's genital slit. Curiously, zebra sharks will sometimes flip onto their backs to receive a belly rub from a diver. This suggests to me that it feels good; otherwise, why go to the trouble or take the risk?

Is there something seedy about presenting photos of animals engaging in sexual intercourse? While I was compiling images for this book, one contributor expressed his concern that a section on sex might be voyeuristic. Though I do not believe animals are shy about sex in the manner of humans, how our reactions affect our attitudes to them does matter. Animals enrich our lives by their presence, and if they did not mate they would soon disappear. For that reason alone, their procreative behavior is as much worth celebrating as any other. I hope you agree.

MASAI GIRAFFE (*Giraffa camelopardalis tippelskirchi*), Keekorok, Masai Mara, Kenya. Giraffes have a promiscuous mating system. Males are largely opportunistic, though they seem to prefer mating with younger females. Females, who are choosy and often play hard to get, tend to favor older, more dominant males. A female may communicate an interest in mating with a male by approaching him and rubbing her neck on his flanks. Photo: Marjorie White.

RING-NECKED PHEASANT (*Phasianus colchicus*), Lee Metcalf National Wildlife Refuge, Montana. A male ring-necked pheasant performs a wing-flapping display in an effort to convince a female to mate with him. Males also dance around their partners with lowered wings in a display called waltzing. Photo: Ken Archer.

BARNACLE GOOSE (*Branta leucopsis*), Oostmaarland, Belgium. In species for which parental cooperation is vital to successful rearing of young, we may expect strong emotional attachments and pleasurable feelings linked to pair maintenance. Barnacle geese mate for life, and research shows they prefer to pair up with familiar individuals from their own region. Here, a pair demonstrates some of the behaviors associated with copulation. Photos: Jonathan Lhoir.

LEOPARD SHARK (*Stegostoma fasciatum*), Andaman Sea, Bay of Bengal, India. Copulation in leopard sharks lasts from two to five minutes and is preceded by a courtship in which the male grasps the female's tail or a fin in his mouth (above) and, if successful, flips her onto her back (left). Photos: Vandit Kalia.

BURGUNDY SNAIL (*Helix pomatia*), Plainevaux, Belgium. Land snails are hermaphroditic, which means an individual can mate with itself and produce fertile offspring. This is a useful adaptation for slow-moving animals, whose chances of encountering another individual may be poor at times. But copulation between two snails is better from a genetic perspective and, perhaps, pleasurable. Studies have shown that snails will actively stimulate parts of their brain associated with reproductive behavior, which suggests that they might be capable of positive sensations. Photo: Jonathan Lhoir.

JAGUAR (*Felis onca*), Cuiaba River, Pantanal, Mato Grosso, Brazil. A pair of jaguars enjoy a quiet moment of intimate coupling. Photos: Mark Andrews.

RED DEER (*Cervus elaphus*), Richmond Park, London, England. As members of harem species—in which the strongest and fittest males amass a herd of does and try to prevent other bucks from mating with them—deer are often portrayed as sexually aggressive males and passive females. This amorous hind mounting a stag shows that the females are not as passive or indifferent as we are often led to believe. Photo: Elliott Neep.

COMMON BLUE BUTTERFLY (*Polyommatus icarus*), Leiden, the Netherlands. Butterfly mating is more complex than it appears and is often preceded by courtship behavior. In some species, males deposit a plug in the female to try to prevent subsequent males from successfully fertilizing her. It is commonly assumed that insects feel nothing, but this is a presumption on our part. Photo: Arthur Sevestre.

HOUSE SPARROW (*Passer domesticus*), Stoney Creek, Ontario, Canada. House sparrows are usually monogamous throughout their lives. To mate, as shown here, the male reaches around and beneath the female's raised tail to bring their reproductive organs into contact. Photo: Raymond Barlow.

Love

L ove is any of various emotions that relate to feelings of strong attachment to another. The origin of such feelings probably lies in their benefit to inclusive fitness—the sum of an organism's reproductive output and that of relatives with shared genes.[1] Love motivates individuals to care for and protect one another, which in turn confers a survival advantage. For example, parents who are emotionally attached to each other are more likely to cooperate effectively in raising young. But while love has origins that may be ultimate and evolutionary in nature, it is also an emotion felt by individuals.

The object of our love may be a mate, a child, a parent, a close friend, or even someone we idolize but have never met. Loving feelings include the passion, intimacy, and desire that accompany romantic love and the nonsexual emotional closeness of familial and platonic love.

On the question of love's existence in the hearts and minds of animals, science has been mainly mute. Few textbooks on animals discuss the possibility of love. For instance, the word *love* can be found in neither the index of *The Oxford Companion to Animal Behavior* nor the *Encyclopedia of Animal Behavior*.[2] There are, I think, two main reasons for this. First, it is difficult, if not impossible, to prove feelings of love in another individual, even a human. This is the challenge of private experiences. It is why the study of animal feelings in general was largely neglected for the century following the 1872 publication of Charles Darwin's *The Expression of the Emotions in Man and Animals*. But humans can

---

Previous spread: **BLACK OYSTERCATCHER** (*Haematopus bachmani*), Ruby Beach, Olympic National Park, Washington. After fledging at about forty days, oystercatcher chicks remain with their parents until the next breeding season and fly with them on migration. Youngsters learn foraging techniques from their parents, who ply them with food during the apprenticeship. Recent studies show birds have strong and lasting emotions, and it seems likely that loving feelings develop between parents and chicks during this prolonged familial period. Photo: Ken Archer.

JAPANESE MACAQUE (*Macaca fuscata*), Detroit Zoo, Michigan. Can we really doubt that a monkey feels love for her baby? Recent studies with baboons show that the loss of an infant triggers the same long-term hormonal changes in the mother that have been documented in women grieving the loss of a child. The monkeys also seek therapy by expanding their social network. Their version of sending flowers and condolences is to spend more time being groomed by and grooming members of their social circle. Photo: Robert Parnell.

at least give verbal expression to their loving feelings; so far, animals cannot, although there is the potential for revelations from language-taught great apes.

Second, our sense of superiority over other animals has made us loath to accept the idea that they can have such presumably complex feelings as love. That nonhumans are conscious remains controversial for some scientists, although their numbers are dwindling. Nevertheless, biologists usually use the term *bond* in place of *love* when referring to nonhumans. This is a safety net to avoid anthropomorphism.

As scientific interest in animal emotions has grown in recent years, new discoveries have suggested that animals too can feel love. One such discovery is that spindle cells occur in nonhumans. These large neurons, named for their shape, occur in parts of the human brain thought to be responsible for social organization, empathy, and intuitions about the feelings of others. Spindle cells are also credited with allowing us to feel love and to suffer emotionally. Long believed to exist only in the brains of humans and other great apes, in 2006 spindle cells were discovered in the same brain areas in humpback whales, fin whales, killer whales, and sperm whales.[3] Furthermore, the proportion of spindle cells in whales' brains is about three times that in human brains. It appears that spindle cells evolved in whales about thirty million years ago, some fifteen million years before humans acquired them, so the fact that the common ancestor of cetaceans and primates lived more than ninety-five million years ago means that spindle cells evolved separately in these lineages.[4] In 2008, spindle cells were reported in both African and Indian elephants.[5]

Despite our similar biology, it doesn't necessarily follow that whales or elephants can feel love in the manner that we can. But we cannot take for granted the complexity of these animals' social behavior. Elephants are more easily studied than whales, and like whales they are long-lived, large brained, and strongly social. They appear to be vulnerable to the same sorts of long-term psychological conditions that may afflict humans who have suffered mental or physical trauma: there is solid evidence emerging that they feel emotions relating to grief at loss and to post-traumatic stress disorder (PTSD).[6] Poaching and so-called elephant culls have left many orphaned elephants in the care of compassionate humans. Growing up with the traumatic memories of terror and the loss of a mother or another close companion, exacerbated by the dearth of nurturing that only a mother can provide, these orphans show the classic symptoms described in human PTSD patients, including sleep disorders, reexperiencing (including what appear to be nightmares), loss of appetite, irritability, and hyperaggression.[7] These are not pleasurable feelings, but brains that are capable of them might be capable of feelings of love too. And needless to say, love isn't all about pleasure. The emotions felt toward a loved one can quickly turn to grief, anger, or resentment, depending on circumstances.

There is also much evidence in elephants for tight emotional bonds between adults—particularly members of the same family group—and between mothers and their calves. The elephant experts Iain Douglas-Hamilton, Cynthia Moss, and Joyce Poole, who have cumulatively studied African elephants for more than one hundred years, have published many accounts of elephants' emotional attachments. For example, when Eleanor, the matriarch of a family unit called the First Ladies, became gravely ill and fell to the ground, she was aided by Grace, the matriarch of another family, called the Virtues. Seeing her down, Grace ran over to Eleanor with her tail raised and temporal glands streaming secretions, sniffed and touched Eleanor with her trunk and foot, then used her tusks to help lift Eleanor to her feet. But the effort was ultimately unsuccessful, and during the week following Eleanor's death, elephants from five family units visited her body.[8] We may wonder whether these behaviors were motivated by loving feelings, but there can be no question that elephants show deep emotional concern for others whom they know.

There are many accounts of other mammals and birds of a variety of species showing the behavioral hallmarks of grief from loss of a loved one: lethargy, disinterest, decreased appetite. The ethologist Konrad Lorenz described the sunken eyes and hunched posture of newly bereft graylag geese: "A greylag goose that has lost its partner shows all the symptoms described in young human children."[9] Jane Goodall has described the "hollow-eyed, gaunt and utterly depressed" state of Flint, an eight-year-old chimpanzee, following the death of his elderly mother, Flo. Flint had been more dependent on his mother than most chimps his age, and the loss of Flo imploded his universe. He stopped eating and died three weeks later, curled up at the spot where he had found Flo's body.[10]

Depending on others is an ingredient in the evolution of love. Loving feelings are especially important for animals who work together to raise young. The biologist Bernd Heinrich, who has studied ravens for many years, says: "I suspect they fall in love like we do, simply because some kind of internal reward is required to maintain a long-term pair bond."[11] The pied-billed grebes pictured on pages 118 and 119 have a lot of work ahead of them: these two birds will share the tasks of nest building, incubation, and looking after their chicks, including letting them ride on their backs. Feelings of love could help to keep them happy and content to face the challenges of child rearing. A pair of grebes who didn't love each other might be less willing to cooperate.

One of the theories for the elaborate courtship behaviors found in many animals is that they help the participants to assess the suitability of potential mates. How fit is she? How responsive? How much does he like me? Two creatures paired in "marital" bliss are more likely than a dissatisfied couple to successfully raise babies to adulthood—and those successful babies will carry the loving genes of their parents.

Sac-winged bats perform elaborate courtship displays with acoustic, visual, and olfactory elements. While females roost on tree trunks, males hover in front of them. Their wing beats fan the females and waft scents from the pouches for which these bats are named. The males also have a repertoire of calls including songs unique to each that are produced exclusively during their courtship flights. It is thought that females gauge the desirability of males by components of their songs, which include trills, noise bursts, "short tonal" calls, and "quasi constant frequency calls." This is a harem species in which some males mate with several females and others mate with none, so perhaps this all-or-nothing competition for females' affections drives the complexity of the songs.[12]

In humans, feelings of love are accompanied by changes in the brain's chemistry. As people fall in love, the brain begins to release the hormones dopamine, norepinephrine, and serotonin. These chemicals stimulate the brain's pleasure centers, creating the rewarding feelings that come with lust and attraction toward another.

Animals show comparable biochemical changes in similar situations. For instance, when a male zebra finch sings a courtship song to a female, nerve cells in a part of his brain called the ventral tegmental area become activated. In humans this is the same area that responds to cocaine, which in turn triggers the release of the brain's "reward" chemical: dopamine. Male zebra finches don't show this brain response if they sing solo; it is only in the presence of a potential mate that they have this pleasurable reaction. This suggests an emotional charge experienced by the male bird—and I suspect by the female also, if she is interested. The study, conducted at the Riken Brain Science Institute in Saitama, Japan, is considered the clearest evidence so far that singing to a female is pleasurable for male birds.[13] There is also evidence that zebra finches find the sight of their mate rewarding. Separated pairs call and search vigorously for their mate.[14] Isolated males will work (having been trained to hop from perch to perch) to be allowed to see female zebra finches through a window, and they will work much harder to see their mate in particular.

Love can also be detected by brain imaging studies. Magnetic resonance imaging of sexually aroused common marmosets—small monkeys that usually form monogamous pairs—shows patterns of brain activation and deactivation that are shared by women experiencing feelings of romantic love.[15] By itself, this provides little support that marmosets experience loving feelings, but scientists could run a more compelling study to measure and compare brain activity in marmosets presented with images of their mated partner and of an unfamiliar marmoset. As yet, no such study has been done.

Might the physical pleasure of touch demonstrated by the allopreening gentoo penguins on page 111 be accompanied by feelings of love and affection? A rewarding relationship tends to be a more committed one, and if they breed successfully, this dedicated pair will pass along their genes for affection and pleasure to the next generation. It's another avian example of how emotions and pleasures—and not just physical features—are grist for evolution's mill.

We might assume that the depth of an emotion is predicted by the intelligence of the animal and by its evolutionary proximity to humans. Not necessarily. Common chimpanzees appear not to fall in love. Because their mating system is notably promiscuous and child-rearing responsibilities lie solely with the mother, there is no premium on males to prove their commitment to females. This is not to say, however, that chimps are devoid of loving capacities. The attachment between a mother and her child is vital to the latter's survival, and love between the two is strongly evident in this species.

You may be surprised to learn that one of the animals most studied in the realms of emotional attachment is a small rodent. The prairie vole, which inhabits the grasslands of central Canada and the United States, is a gray-brown mouselike mammal with small ears, a short tail, and a yellowish belly. Male and female prairie voles form lifelong pair bonds, huddle and groom each other, and share nesting and pup-raising responsibilities. Like humans, they tend to remain with a single mate, although (again like us) they are not always sexually faithful. Their brains operate in much the same way ours do when it comes to sex and attachment. For example, in both species the hormone oxytocin modulates maternal bonding (loving) behaviors, while vasopressin accomplishes the task in males. Once again, our friend dopamine is part of the mix, regulated by oxytocin. Sometimes referred to as "the cuddle hormone," oxytocin facilitates labor, childbirth and breastfeeding, orgasm, social recognition, and pair bonding. You will not find love ascribed to prairie voles in the scientific literature; instead you will see words like *bonding* and *attachment*. But the hormones are exactly the same in a human and a vole, and the evolutionary benefits align.

Perhaps my favorite image in this chapter is Ken Archer's photo of a pair of beavers enjoying some intimate moments together (page 112). Beaver pairs remain together for several years and sometimes for life, and their family groups can include not only recently born kits but also yearlings from the previous litter. It is hard to know what emotions these long-lived rodents may feel for one another. But as their relationships endure, it would be unfair to deny them what Colonel E. B. Hamley referred to in *Our Poor Relations* as "the ripening warmth of intimacy."

GENTOO PENGUIN (*Pygoscelis papua*), Cuverville Island, Antarctic Peninsula. Gentoo penguins work closely together to build their nests and successfully rear chicks. Among the behaviors these birds engage in are gifting each other with the much-sought-after stones used to build their nests. Here, a mated pair engage in a bout of allopreening. The pleasure of this sort of physical contact helps reinforce the intimacy of their relationship. Photo: Thomas D. Mangelsen.

BEAVER (*Castor canadensis*), Indian Creek, Yellowstone National Park, Wyoming. Successfully rearing young beavers requires the work of both parents. Species like this are more apt to mate for life, and beavers are no exception. Rewarding behaviors, such as mutual grooming, help to reinforce the feelings each mate has for the other. Photo: Ken Archer.

ATLANTIC GRAY SEAL (*Halichoerus grypus*), Donna Nook, United Kingdom. We can only wonder what this male and female gray seal were thinking and feeling while they frolicked in the shallow surf. Both participants repeatedly approached the other, and the female did not try to escape the male's amorous-looking attentions. At times things got a bit rough, but seals have thick skin, and the overall impression was one of tenderness. Photo: Arthur Sevestre.

WESTERN LOWLAND GORILLA (*Gorilla gorilla gorilla*), Bronx Zoo, New York. Four-year-old lowland gorilla Zola is cradled by his mother, twelve-year-old Tuti. Gorilla infants are dependent on their mothers for up to five years. As for all great ape infants, including human ones, their mother is the center of their universe as they gradually gain social competence and independence. Photo: Vicki Puluso.

PRAIRIE DOG (*Cynomys ludovicianus*), Denver, Colorado. Highly social, prairie dogs are also intensely affectionate toward family and other close members of their colony. Greeting behavior can be ecstatic. Here a mother is surrounded by at least five pups, some of which may not be her own. The pup on the left is doing a "jump-yip" call, which may function as an all-clear signal and is so explosive that it sometimes causes the caller to fall over backward. One or two of the pups in the background are "greet-kissing" the mom by pressing their mouths to hers. Photo: Sandy Nervig.

WILD BOAR (*Sus scrofa*), Plainevaux, Belgium. A wild boar sow follows her babies as they explore a Belgian woodland. Intelligent and emotional, these animals are fiercely protective of their adorable piglets. Photo: Jonathan Lhoir.

ASIATIC LION (*Panthera leo persica*), Gir Forest National Park, India. The last remaining population of Asiatic lions—numbering 411 individuals in April 2010—inhabits a 558-square-mile sanctuary in western India. They face many threats, including habitat loss, falling into and drowning in open wells, and traffic on major roads cutting through the park. Photographer Steve Mandel has established the Lions of Gir Foundation to help rescue these regal animals from the brink of extinction. Photo: Steve Mandel.

GIRAFFE (*Giraffa camelopardalis*), Bronx Zoo, New York. A pair of giraffes show their love for a calf. Photo: Vicki Puluso.

PIED-BILLED GREBE (*Podilymbus podiceps*), Fort Steilacoom County Park, Lakewood, Washington. A pair of pied-billed grebes engage in animated courtship prior to mating. The male, on the right in both images, is lavishing attention on the female. In the second image he can be seen with nesting material in his mouth, which he's showing to the female to get her interested in mating and sharing a nest with him. Courtship allows partners to gauge compatibility and assess suitability as mates. In human terms we also refer to this as dating. Photos: Nate Chappell.

RAINBOW LORIKEET (*Trichoglossus haematodus*), Jurong Bird Park, Singapore. Billing, or kissing, is practiced by many birds. It is an expression of the devotion a mated pair feel for each other. Photo: Melanie Votaw.

BLACK-HEADED GROSBEAK (*Pheucticus melanocephalus*), Madera Canyon, Arizona. Offering food to a prospective mate forms part of the courtship behavior of many birds. This male (on the right) is passing a morsel to the female, who enthusiastically accepts. Both mated birds will share incubation and feeding of the chicks. Unusually, in grosbeaks both sexes sing songs, sometimes while sitting on the nest. Photo: Nate Chappell.

Comfort

Comfort is something we usually associate with leisure or luxury. Settling into a soft chair, pressing one's face into a hot face towel, or bedding down for a good night's sleep—these things feel good. It's as if our bodies were saying: "Ah, that's better."

Actually, feelings of comfort are the product of one of nature's most basic and important requirements: homeostasis. Homeostasis is about maintaining a stable, constant internal condition. If an animal feels too cold, she seeks warmth. Too hot, and she will act to make herself cooler.

Hunger is relieved by the pleasure of a meal, thirst by the refreshment of a drink, and fatigue by the reward of rest. The related pleasant sensations are nature's way of rewarding adaptive behavior—actions that keep us healthy and alert.

It follows that a single stimulus may feel pleasant in one situation and unpleasant in another. The physiologist Michel Cabanac, whom we met in the introduction, has coined the term *alliesthesia* to describe this phenomenon. Temperature offers a good example. Imagine slipping into a warm bath on a cold winter day. It feels good. Now imagine slipping into the same tub having just emerged from a sauna. Most of us would find this uncomfortable. Cabanac had people rate the pleasantness or unpleasantness of a bowl of water at a preset temperature and found the ratings diverged according to the initial temperature of the subject.[1] This is a quantitative way of demonstrating the importance of maintaining homeostasis. And it's why the squirrel on page 134 is probably enjoying the feel of the sun-warmed bricks on a cool day.

Scientists use the technical term *behavioral thermoregulation* to describe actions that help an animal maintain a core body temperature that promotes survival.[2] For example, reptiles often become

---

Previous spread: DOMESTIC "BENGAL" CAT (cross between *Felis catus domesticus* and Asian leopard cat [*Prionailurus bengalensis*]), Fort Myers, Florida. These Bengal kittens had finished nursing a few minutes earlier and then, like just-fed human infants, lapsed into a luxurious, full-bellied sleep. Photo: Buck Ward.

AFRICAN LION (*Panthera leo*), Lion Sands Private Reserve, Sabi Sands, South Africa. This lioness had just gotten up after a long snooze. Like their domestic cat cousins, lions are masters at stretching. The lioness's facial expression captures the distinctive pleasure this activity provides. Photo: Mike Moss.

sluggish during cooler nights, and they must warm up in the morning to pursue daily survival activities that require speed and agility. Moving to a warm place, such as a sunlit rock, allows them to recharge their engines by using solar energy instead of consuming metabolic energy (although the latter allows an organism to live in a broader range of habitats, it is more energetically expensive). Similarly, animals may seek shade or the cooling effects of water when they become overheated. If a cool or a warm spot is not available, they may use other, autonomous means of adjusting their body temperature. These include sweating, shivering, panting, and wing fanning, which, once again, are energetically more costly but can avert hypo- or hyperthermia.

Animals have many techniques for getting comfortable. Mice, meerkats, and other mammals form dynamic huddles on chilly nights; they amass together in a furry pile, sharing one another's body warmth. The huddles are dynamic because individuals on the wind side will periodically move to the lee side. Like mammals, some birds huddle together in frigid weather; I've watched flocks of sanderlings and red knots clustered close on coastal strands, one leg tucked in against the wind. Birds also make admirable nests that to varying degrees provide the comfort of warmth and insulation from harsh elements. Birds of a feather may even congregate in an abandoned nest to keep warm on cold nights. Glenn E. Walsberg at Arizona State University watched fifteen black-tailed gnatcatchers return each day to an abandoned verdin's nest, where they stayed in a communal huddle for five cold December nights in 1989.[3]

On a sunny afternoon in late May I saw a blue jay who, to the uninitiated, might have looked either seriously distressed or deranged. Normally so prim and symmetrical in their mien, birds do not typically adopt the posture this one was in. His body was askew, and one wing was fanned out as if broken. His head was tilted to the side, and his bill was agape. He looked as if he had just survived a run-in with a cat or a windshield.

In fact, he was having a good sunbath. It's something that jays and other birds do that reminds us of our own behavior: they stretch out a wing, fan their tail feathers, and soak up those rays. The jay's slanted aspect placed the larger surfaces of his body at a 90-degree angle to the sun. I've seen starlings and pigeons also performing this comfort behavior. Likewise, the tufted titmice on page 136 have spread their wings and tails and fluffed out the feathers on their backs to optimize their enjoyment of the sun's warmth.

The yin-yang of comfort is that it often involves the relief of discomfort. Slipping into a hot shower feels better if we've been chilly. There is little compulsion to scratch if there is no itch. The young great blue heron on pages 130–131 uses his long feet to attend to an irritation on his neck. Perhaps it is caused by a mite or another of the external parasites that commonly occur on birds. But whatever the cause of the itch, scratching brings relief, and it feels good. Animals with horns or antlers have ready-made tools for self-scratching (see the goat on pages 138–139).

If you don't have horns, perhaps a trunk or even a tail will suffice. At a sanctuary in India, I watched one of the elephants use her coiled trunk to rub an itching eye. Elephant tails are also remarkably versatile, with a network of coordinated muscles running their entire length. I watched in amazement as another elephant lifted and then coiled her tail, pressing the bottom part of the coil she'd made against the tail base before swiping the distal portion of the tail across her anal area. She did this several times after defecating. She also dragged her tail across the folds between her hind legs and belly, perhaps to scratch. On a guided walk in South Africa's Kruger National Park in May 2008, I passed a couple of stumps and tree limbs that had been worn to rounded nubs by countless animals seeking the relief of an itch or the simple pleasure of a good rub. Landmarks like this may also serve the dual purpose of leaving a scent mark, a sort of calling card that tells other animals *I live here*.

Stretching is another comfort behavior many animals engage in. It boosts circulation and helps prepare joints and muscles for action following rest. (If you engage in sports you'll surely know the value of this.) These benefits explain why stretching feels good. Many birds perform a quite stereo-typed behavior of simultaneously extending a wing and a leg backward and away from the body (see the rail on page 137). Cats have various stretching postures, and if you know cat behavior you may recognize a familiar expression on the face of the lion having a stretch on page 126. Cats typically stretch their front legs by extending them straight forward; the hind legs remain upright and the back slopes in a concave arch. Following the foreleg stretch, a cat will often step forward, then lift and stretch each hind leg in succession. Those who enjoy yoga may think of it as the Sun Salutation for cats. The ground squirrel on page 141 also appears to have consulted a yoga manual.

Achieving comfort can be a social activity, as shown by the kittens snuggling after a feeding on pages 124–125 or a tight flock of sandpipers soaking in the rays of the late afternoon sun. A narrow scientific interpretation of the sandpipers' behavior would see their gregariousness only as a means of safety; when you're part of a group, danger is more likely to be spotted sooner. But feeling safe may also contribute to their feeling good. Anxiety is not conducive to comfort or joy. For example, rats who have learned to come running to be tickled or petted by a trusted human companion will not do so if they are exposed to bright lights or the smell of a cat.[4] Other bodies close by can also buffer the wind, a benefit that may help explain why birds standing in a cluster often shift their position.

Christopher Hogwood, the enormous pig who gained fame as the subject of Sy Montgomery's best-selling book *The Good Good Pig*, seemed to enjoy the presence of two familiar young boys who would take turns sprawling atop his flank. Montgomery adopted Hogwood—a runt deemed too sickly to live—from a local piggery. He survived, amassed a vast slops empire, and grew to 750 pounds.[5] Hogwood also grew to love people and welcomed their attentions. Montgomery told me about the relationship between the pig and the boys:

> Ned Rodat, then 4, and his older brother, Jack, visited Christopher regularly. The kids' parents would save their leftover pancakes and stale bagels in the freezer in Boston all week so they could bring them up to Chris on the weekends. I have no doubt that if Christopher had been unhappy about the kids lying on him, he would have let us know! Pigs aren't shy when it comes to demonstrating how they feel. Christopher loved having Ned on him, and Jack, too.[6]

It appears that Christopher Hogwood enjoyed both the camaraderie and the contact. I find something deeply satisfying and—dare I say—comforting about the weight of one of my cats resting on my lap or my back while I read or nap, and I think it's reasonable that a pig can feel that way too.

GREAT BLUE HERON (*Ardea herodias*), San Carlos Bay, Florida. This juvenile heron has yet to attain the characteristic adult plumage of her species. Perhaps what's causing the itch is a new feather beginning to emerge. Her long foot brings the pleasure of relief. Photo: Buck Ward.

SILKY SIFAKA (*Propithecus candidus*), Berenty Reserve, Madagascar. Wouldn't it be nice if we could remember that period of early infancy when all our needs were taken care of? At left, an infant sifaka (a species of lemur) nestles cozily in the dense fur of his mother's belly. Photo: Steve Mandel.

JAPANESE MACAQUE (*Macaca fuscata*), Jigokudani Monkey Park, Nagano, Japan. Mother macaques are highly nurturing of their children, in whom they invest prolonged care. This youngster is lapsing into a blissful sleep in the world's safest and coziest place. Photo: Andrew Forsyth.

WESTERN GRAY SQUIRREL (*Sciurus griseus*), Hawthorne, California. Many mammals and reptiles use the heat absorbed by rocks—or bricks—to warm up. It was cool out when the photographer captured this squirrel enjoying the radiating warmth of bricks in the midday sun. The rodent has flattened his belly to maximize the pleasurable transfer of heat. Squirrels use the same technique to cool down when a surface is colder than the ambient temperature. Photo: Mike Moss.

GRIZZLY BEAR (*Ursus horribilis*), Khutzeymateen Grizzly Bear Sanctuary, British Columbia, Canada. Bears have few enemies (save humans) in their natural habitats, so they feel little need to seek out a sheltered place for their siestas. This bear has found the perfect grass-lined perch for a midmorning snooze. Photo: Dave Hutchison.

TUFTED TITMOUSE (*Baeolophus bicolor*), Weymouth, Massachusetts. At first glance we might think these birds are in distress. One of the challenges of appreciating animals' pleasure is to understand their behavior in context. By spreading their wings and tail feathers, these titmice create a larger surface area for soaking in the sun's warmth. Fluffing up their feathers also allows the sun's rays better access to their bodies. Photo: Kathy Vespaziani.

CARRION CROW (*Corvus corone*), Plainevaux, Belgium. Birds often bathe to cool off on a hot day. Maintaining a stable body temperature is vital to survival. It's why something cool feels good when we are hot and vice versa (see the discussion of alliesthesia on page 127). Photo: Jonathan Lhoir.

VIRGINIA RAIL (*Rallus limicola*), Malheur National Wildlife Refuge, Oregon. It behooves a bird to be ever ready for flight. To stay limber, birds often perform wing stretches, fully extending each, usually one after the other and often accompanied by a leg stretch. Wings need to be ready for quick deployment in an emergency, so the behavior probably serves as a preparedness strategy. It also likely feels good, which encourages the birds to keep doing it. Photo: Ken Archer.

DOMESTIC GOAT (*Capra aegagrus*), Farm
Sanctuary, Orland, California. Goats
and other horned animals have built-in
back scratchers. Photo: Connie Pugh.

RING-TAILED LEMUR (*Lemur catta*), Berenty Reserve, Madagascar. The sun is the energy foundation for practically all life on earth. And, as any sunbather knows, it feels good, especially when you're chilly. Photo: Steve Mandel.

ARCTIC GROUND SQUIRREL (*Spermophilus parryii*), Denali National Park, Alaska. Stretching is important to boost the blood supply to muscles and joints before they spring into action. Not surprisingly, then, it feels nice. Photo: Anthony Gibson.

CALIFORNIA QUAIL (*Callipepla californica*), Rancho San Antonio Open Space Preserve, Los Altos, California. Dustbathing helps to keep feathers clean and dry and may repel parasites. Birds are strongly motivated to perform this activity, and it brings satisfaction. Photo: Connie Pugh.

Companionship

Even the most solitary of animals must at some point consort with members of their own species—that is, if they expect to reproduce. Among the more social species, companionship is part of everyday life.

Social living is a triumph of cooperation over competition. Companionship has numerous benefits. Congregating with others permits the sharing of information, such as where to find food or a good roosting site. Groups may also be able to secure food that a single individual might not. The success of lions, hyenas, and wolves lies in their ability to subdue large prey that an individual alone could not defeat. Individuals may also offer help to another in danger or duress. People as far back as the ancient Greeks have described this behavior in dolphins, which have been observed saving other dolphins by biting through harpoon lines, supporting sick companions near the surface to keep them from drowning, and staying close to females in labor. Whales have also placed themselves between a hunter's boat and an injured comrade.

Being helpful or otherwise showing kindness to others feels good. We know this from our own experience. And we too are subjects of nature's system of rewards and punishments. It's no coincidence that kindness feels good, for it benefits us too.

Another benefit of companionship is increased vigilance for potential danger. If you look at a flock of geese foraging in a field, you will almost certainly see at least one of them acting as a sentry, neck stretched upward and eyes alert to any possible danger. Sentry duty is an important safety

Previous spread: DIAGONAL-BANDED SWEETLIPS (*Plectorhinchus lineatus*), Great Barrier Reef, Queensland, Australia. We tend to look at fishes shallowly, as anonymous members of a species. Schools certainly have adaptive benefits, but they are more than mindless aggregations. New research shows that fishes are individuals with distinctive personalities. They recognize others and develop preferences for whom they like to swim with. Photo: Fred Bavendam / Minden Pictures.

DOMESTIC CATTLE (*Bos taurus*), Cypress Hills, Saskatchewan, Canada. Cattle also have their favorites. The calf on the left has just walked up to the other and is expressing affection. Immediately following the photo, the white calf moved her head up the side of the brown calf's head, nuzzling the other's cheek and both ears. She then rested her chin on top of the other's head, closing her eyes for a few moments and looking blissful. Then she lay down and they snoozed together. Photo: Rosanne Tackaberry.

measure while other flock members have their heads bent down at grass level. The birds switch roles so everyone gets time to feed in relative peace. Foraging in larger groups has been shown to reduce individual vigilance in many animals, including wild boars, kangaroos, and ostriches.[1]

If a vigilant forager sees something alarming, he or she must be able to alert the rest of the group. Prairie dogs use specific calls for different predators. The call for an aerial predator sends all nearby colony members diving down the nearest burrow, a coyote alarm call signals them to watch carefully from a burrow entrance, and if the enemy is a domestic dog they may just stop what they're doing and look about alertly.[2] Such is the degree to which prairie dogs have been persecuted by hunters that some populations even modify their calls to indicate that a person is carrying a gun.[3]

Animals also use calls to convey less-troubling information. Chickens respond flexibly to the calls of others. For instance, a hen who already knows there is no food present will show less interest in a rooster's food-solicitation call than will a naïve hen. Other animals now known to benefit from the representational calls of their companions include monkeys, tamarins, lemurs, warblers, chickadees, quail, and meerkats. Because these calls are distinctive, refer to specific threats, and are interpreted appropriately by listeners, some scientists have begun to refer to them as words.[4]

Call interactions are not always confined to members of the same species. Hoofed mammals respond to the alarm chatters of the oxpecker birds who glean parasites from their hides.[5] Many birds forage in mixed-species flocks, benefiting from the collective vigilance of others, which requires knowing their lingo.

It used to be thought that a wild animal's only important companions were close relatives, but that notion has been corrected. Nature is full of cooperating animals who are not close relatives. In evening bats, for example, about 20 percent of nursings involve females feeding unrelated pups.[6] One reason for such seemingly selfless behavior is that other animals reciprocate. They remember and appreciate good deeds done for them, and they are likely to return the favor.

The evolutionary upshot of all these benefits of hanging around with others is that nature should reward it. And she does. As a highly social species, humans know the good feelings that come from being with others. Companionship can provide comfort, security, intimacy, solidarity, and a sense of belonging, to name a few. Another benefit of companionship is the simple joy of playful interaction. As we saw in the first chapter, on play, animals spend time cavorting and romping.

Rewarding or not, animal feelings may be expressed in surprising, unexpected ways. I have no earthly idea what the wild mule deer and the housecat Gizmo were thinking when they connected on Becky Hermanson's porch in Baldwin, North Dakota (see page 161). But connect they did, and Becky's photo gives no hint of fear or apprehension between them. One could be forgiven for thinking they were old friends, but that seems unlikely. I asked Becky about their encounter, and here is her reply:

> The mule deer was chased out of my pasture by my horses and jumped the fence and came into my yard, where I was with my granddaughter. For some reason she loved my cat. I thought maybe the deer had been someone's pet. I called the Game and Fish Dept. They said that mule deer are not very common in my area, and they thought she was just looking for companionship of some sort. She

stayed around for a few hours, then left after some encouragement from me, as I did not want her to be a pet. She belonged in the wild.

I find the cat's behavior at least as remarkable as the deer's. It seems Gizmo was able to read the benign, friendly intent of her distant mammalian cousin.

Several photos here show birds perching together. The two redstart fledglings on page 155 were probably huddling together to stay warm. The pair of carmine bee-eaters on page 156 and the dusky woodswallows on page 154 belong to species known for their gregariousness. Carmine bee-eaters are believed to be monogamous, and the males sometimes give an insect gift to their mates following copulation. Woodswallows perch notably more compactly than most other gregarious birds, such as starlings; instead of maintaining an orderly distance between neighbors, they press up against one another.

Contrary to the impression of random association that a flock of starlings or a herd of impala may give, animals have favorites among their kind—friends with whom they prefer to spend time and share comforts. We tend not to hold vultures in as high regard as we do cute little songbirds, but they are no less worthy of being, and I'm delighted to include an image of them in this book (see the griffon vultures on page 66). On a visit to Mexico in 2008, as I stood beneath an earpod tree (its burgundy seedpods are shaped like a human ear), a vulture came in to land. It was late dusk, and the bird was preparing to roost for the night. A minute later a second vulture alit in the same tree. Each hopped about on different branches for a couple of minutes, as if gauging them for suitability. They had hundreds of feet of branches to choose from, yet when they finally settled down they were just a few inches apart. Then they alternately preened themselves and each other. Black vultures pair for life, and I surmise that this was a mated couple. The sinister mythology attached to these scavengers doesn't evoke such tenderness. Who knows what other secrets these birds keep from us? We can say all we like about the evolutionary benefits of pair bonding, but we must also be cognizant that animals experience their lives. Vultures have minds and emotions, and they live for decades. It is considered unscientific to say that they enjoy each other's company and to suggest that they may fall in love. I think it is short-sighted and cynical to pretend they don't experience these feelings.

Like vultures, cattle seek out one another's company. A herd isn't just a random collection of individuals wandering about and grazing. Cows recognize individuals and have pals, usually numbering two to four special favorites with whom they spend most of their time, often grooming and licking one another.[7] You may be surprised to learn that fishes are also companionable. A school of fishes swimming in unison may give the impression that there's no selectivity in their associating behavior, but fishes do in fact have shoalmates. Studies of guppies and minnows, for instance, show that schools are by no means composed of random groups of individuals and that these little fishes are quite choosy in whom they hang out with.[8]

Companionship is not, of course, all about pleasure. The wages of friendship include the emotional tug of loss. Following the death of an infant, mother baboons show higher levels of glucocorticoid hormones, which are also associated with grief in women who have lost a baby. The mother baboon's closest companions also show elevated hormones, with the highest levels occurring in individuals who are most closely related to the infant. They feel the pain of loss the most. Bereft mother baboons respond—as do women—by expanding their social networks, which for baboons involves increased rates of grooming.[9] This appears to be a form of therapy to help cope with their bereavement, which lasts a month or more.[10]

Social living in animals has also led to feelings such as empathy, of which there are now countless examples in the scientific literature.[11] Chimpanzees who witness aggressive interactions contact the victims more often than they contact the aggressors.[12] A study of mice found that they became more sensitive to painful events when they could observe another mouse in pain—but only if the other mouse was a familiar individual.[13] Interestingly, this response—described by the researchers as a form of "primitive empathy"—was only shown by female mice, and not males. The scientists who conducted the study injected some mice with painful acids, causing them to writhe in pain. Other mice were treated with chemicals to make them deaf or unable to smell. As with virtually all laboratory rodent studies, the mice were killed following the experiments. We might ask whose empathy toward the mice was more developed: theirs or ours.

BARBARAY MACAQUE (*Macaca sylvanus*), Emmen, the Netherlands. These macaques are part of a fairly large captive group. While many of the others were actively playing and foraging in their enclosure, this trio preferred to relax in one another's company. Here, the female pauses between grooming bouts to press her tongue affectionately against the older male's chin, while the youngster enjoys his cradling warmth on a cold morning. Photo: Arthur Sevestre.

DOMESTIC PIG (*Sus scrofa domestica*), Farm Sanctuary, Orland, California. These are two of seven piglets born to Rosebud, a sow rescued from the Iowa floods of 2008. She birthed them at a farmed-animal sanctuary, and they were the first babies she was allowed to keep. One of the playful, well-loved piglets' favorite things to do was to untie visitors' shoelaces. The family now lives at Lighthouse Farm Animal Sanctuary in Oregon. Photo: Wanda Embar.

DUSKY WOODSWALLOW (*Artamus cyanopterus*), Canberra, Australia. Endemic to Australia, dusky woodswallows are social and gregarious. Day or night, they may be seen roosting communally in numbers sometimes exceeding one hundred. Energy conservation and predation detection are among the adaptive benefits of social roosting. Concomitantly, the birds probably feel more secure. Photo: Simon Bennett.

BLACK REDSTART (*Phoenicurus ochruros*), Buchs, St. Gallen, Switzerland. The photographer found these two newly fledged redstarts on a cold, windy, and rainy afternoon, huddled together on the entrance step to a horse stable. The roof provided them with some protection from the rain but not the cold. They are insulating themselves by fluffing up their feathers and sharing body warmth. It reminds me of huddling next to a space heater on a winter morning when I was a small child. The parents were observed feeding their young on several occasions during the following days. Photo: Dennis Lorenz.

CARMINE BEE-EATER (*Merops nubicoides*), Okavango River, Botswana. Carmine bee-eaters are very gregarious. A single breeding colony can have as many as ten thousand nests and remain in place for eighty years or more. Family members often perch together, and pairs engage in courtship feeding (see pages 60–61). Photo: Tom D. Mangelsen.

EASTERN GRAY SQUIRREL (*Sciurus carolinensis*), Montreal, Quebec, Canada. This trio of half-pint siblings benefits from staying together as they learn the ropes of life outside the nest. Extra eyes, ears, and noses enhance danger detection. Warm bodies help mitigate a cold morning. And there's a ready supply of playmates. Photo: Robert Ganz.

SEA OTTER (*Enhydra lutra*), Pacific Coast, North America. Sea otters often wrap strands of kelp around their bodies to act as moorings as they nap on the water surface. They use a similar technique—linking paws—to stay close to their buddies. Photo: Bob Bennett/Minden Pictures.

NORWAY RAT (*Rattus norvegicus*), Tucson, Arizona. Three young rats snuggle in a towel. From the left, they are Delphiniums Blue (see page 77) and his brothers Cecil and Sorren. They spent their lives together in a loving home and were quite bonded to one another. Photo: Brandi Saxton.

MOUNTAIN GOAT (*Oreamnos americanus*), Mount Evans, Colorado. Mountain goat nannies usually have only one kid at a time, so these three are almost certainly not siblings. But as with the squirrels on page 157, their lives are enhanced by playmates. They are born precocial (eyes open; alert and ready to move about), which is especially useful when you live on steep mountains. Nevertheless, nannies sometimes position themselves just below youngsters to prevent free falls. Photo: Allen Thornton.

MULE DEER (*Odocoileus hemionus*) and **DOMESTIC CAT** (*Felis catus*), Baldwin, North Dakota. After being chased out of a horse pasture, this wild mule deer took a shine to Becky Hermanson's cat, who didn't seem to mind the attention. The deer was quite tame and perhaps had been reared by humans. After a few hours, she returned to more natural haunts. Photo: Becky Hermanson.

DOMESTIC GOAT (*Capra aegagrus*), Farm Sanctuary, Orland, California. Goats are affiliative, tactile animals; they have their best buddies in the flock, and they express their preferences with gentle contact. Here, Hal (on the left) uses a soft nuzzle to show Goddall that they are friends. They were among twelve goats—all very ill, malnourished, and with overgrown, rotted hoofs—rescued in May 2008 along with a cow and a sheep from a slaughter facility in Watsonville, California. Photo: Connie Pugh.

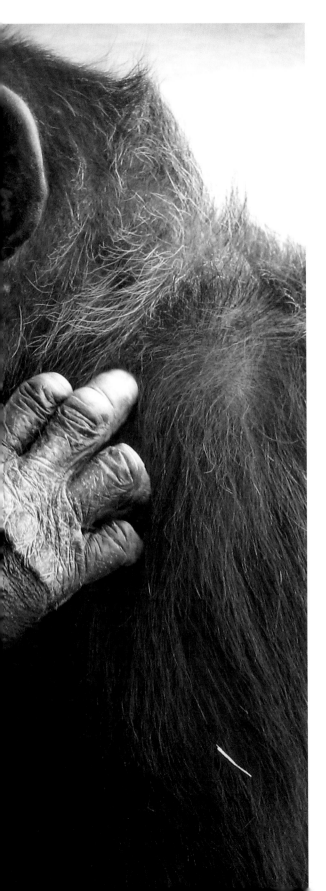

COMMON CHIMPANZEE (*Pan troglodytes*), Chimp
Haven Inc., Keithville, Louisiana. Chimpanzees are as social as we are. They don't merely
want companionship—they need it for normal
psychological health and development. Here,
Teresa and Sheila, two of the older citizens
at Chimp Haven, a sanctuary for individuals
rescued or retired from laboratory research
facilities, groom each other intently. Grooming
is a social lubricant; it cements friendships,
reconciles disputes, and tells another "I care
about you." A large part of its power is that it
feels good. Photo: Amy Fultz / Chimp Haven, Inc.

Other Pleasures

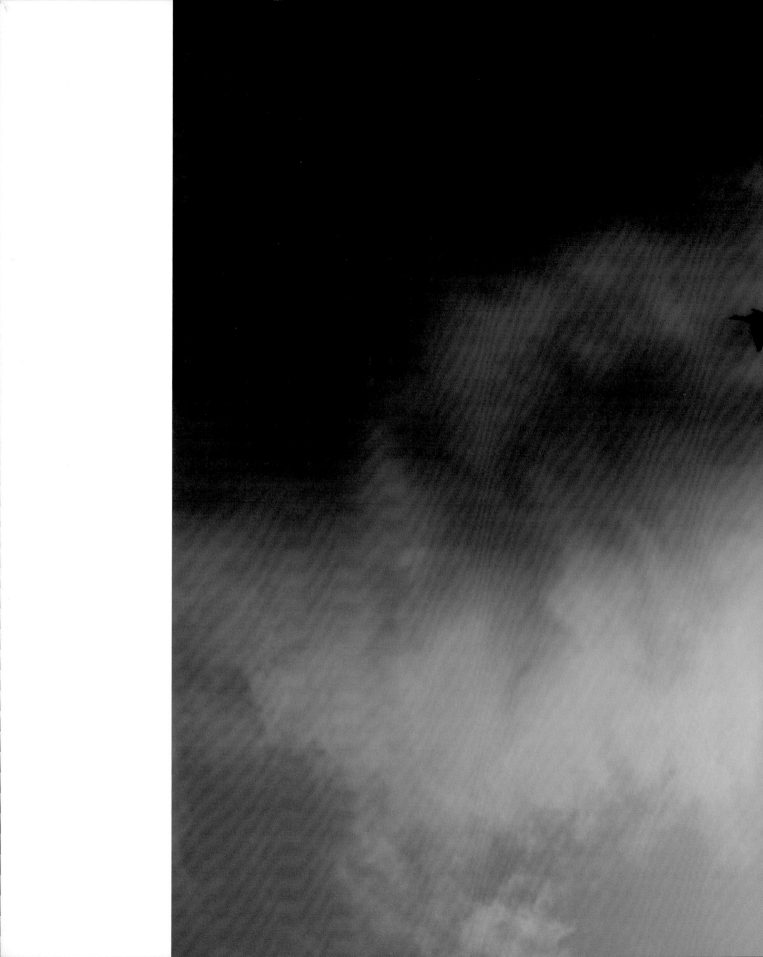

The sources of pleasure so far covered—play, food, touch, courtship and sex, love, comfort, and companionship—are not exhaustive. There are many other experiences that may inspire good feelings. Animals might, for instance, take pleasure in aesthetic beauty. Some might be thrill seekers. Others might have a sense of humor or mischief. We might include curiosity on the list of emotions that may be a source of animal pleasure.

Evolution has forged diverse paths to aesthetic beauty. Flowers and fruits evolved specifically to attract animals, whose ability to move from one place to another facilitates the dispersal of a plant's seeds, which otherwise must compete with (and likely lose to) the parent plant for sunlight, water, and soil nutrients. Bright colors, alluring smells, and sweet tastes are among the sensory rewards on offer to the frugivorous birds, bats, and other animals who partake. The enormous diversity of flowering and fruiting plants on earth today—as many as four hundred thousand species—attests to the power of aesthetic attraction in nature. As Michael Pollan writes in *The Botany of Desire*, "As parties in this grand evolutionary bargain, animals with the strongest predilection for sweetness and plants offering the biggest, sweetest fruits prospered together and multiplied, evolving into other species we see, and are, today. . . . Desire is built into the very nature and purpose of fruit."[1]

Part of what makes a flower attractive might be its symmetry, which may be an indicator of evolutionary fitness. A study of capuchin and squirrel monkeys showed that they prefer regular and symmetrical visual patterns to irregular and asymmetrical ones.[2] In another study, chickens' preferences

Previous spread: **SNOW GOOSE** (*Chen caerulescens*), Freezeout Lake, Montana. A flock of geese slices through a dusk sky: freedom on thirty wings. Photo: Ken Archer.

LONGTAIL WEASEL (*Mustela frenata*), Malheur National Wildlife Refuge, Oregon. Curiosity is very useful if you're a predator. Weasels are always on the lookout for an opportunity for their next meal, but they also take time to investigate other things of interest, such as a curious photographer. When this one first saw the human, she (or he) ran away but kept turning around and looking back at him. Eventually, the weasel returned and peeked over a rock for a closer look. Photo: Nate Chappell.

for particular human faces showed an uncanny correspondence of 98 percent with those of human raters.[3] It is known that humans find symmetrical faces more attractive than asymmetrical ones.[4] It seems unlikely that the chickens are making an aesthetic evaluation of human attractiveness, per se, but they evidently have a keen perception of symmetry.

Despite the agreement of chicken and human subjects in this study, we should be aware that a given animal's perception of what is or is not beautiful may differ greatly from our own. We can be fairly confident that our sensory response to a sweet fruit is similar to a monkey's or a fruit bat's. But I wouldn't want to bet that a rotting carcass has no aesthetic interest for a vulture or that a female flounder fails to appeal to her mate (or a male his).

Sex is another driver of beauty. Many birds are especially attractive to our senses, but their loveliness didn't evolve for our benefit. The extravagance of a pheasant's plumage or a lark's courtship song—these are believed to be the result of choosy females. Studies have shown that females of many species preferentially mate with showier males. One theory is that males with more elaborate adornments may be more fit in the evolutionary sense, and therefore their genes will tend to produce more robust offspring. In fact, the sexy son hypothesis, first proposed in 1979, posits that females (of some species) choose mates solely on the strength of their physical attractiveness.[5] The idea is that if more-attractive males get more matings than less-attractive males, then females who mate with them will tend to produce attractive male offspring, who in turn will leave proportionately more grandchildren carrying the female's genes. The sexy son hypothesis is one of several possible explanations for the evolution of animal beauty and for the diverse and often astonishing ornamentation displayed by some animals.

A skeptic could argue that in choosing more-attractive males, females are merely making a discrimination without any associated positive experience and that therefore there need not be any feelings of pleasure involved. Such a view assumes that birds cannot think or feel emotions. But there is ample evidence that birds are cognitive, and there is a plethora of anecdotal indications that they are emotional.[6] And while very few scientists have quantitatively explored emotionality in birds, that too is beginning to change.[7]

We may rightly wonder whether an insect can take any pleasure in aesthetic beauty, but one of the collateral benefits of their evolutionary pathways is that we may appreciate their beauty. It is surely more than coincidence that we can find beauty in so much of what nature has to offer. After all, humans are animals too (in the biological rather than the semantic sense). We evolved in natural settings. Our perceptions of nature are woven deeply into our aesthetic consciousness. We may have come to appreciate bright, warm colors and round objects because in nature these denote ripe fruit—and the sun. The symmetry of our architecture and technology resonates with the symmetry of animals. Most of the beautiful creatures featured in this book are larger ones. But there are many more tiny creatures that, although usually overlooked, have a beauty all their own when examined closely. The metallic blue sheen of a beetle's elytra (wing covers), its dimples spangling in the light, or the jeweled eye of a tokay gecko may make one wonder why nature would go to so much aesthetic extravagance. The yellow weevil photographed by Andrey Antov is a case in point (page 174). If this image moves you as it did me, I encourage you to get hold of Piotr Naskrecki's book *The Smaller Majority*. It's a visual feast of mostly insects but also many other mini-beasts.

Who is there in the tiny world of an insect to appreciate their color or symmetry? Evolutionary explanations for such beauty usually revolve around what it communicates to other, larger animals. Conspicuous bright colors and bold patterns often signal to would-be predators that an animal is toxic (e.g., poison-arrow frogs), foul tasting or smelling (skunks), or painful (wasps and hornets) and should be avoided. Other nonnoxious animals may benefit by mimicking the model's striking appearance—the palatable viceroy butterfly's resemblance to the poisonous monarch butterfly is perhaps the best-known case, but there are innumerable others. These are examples of how nature can encourage what to some animals is beautiful. The question of whether other animals might perceive as beautiful things that we perceive as beautiful hasn't been put to scientific scrutiny. I think it should be.

Thrill seeking is well known in human circles but little studied in nonhumans. Many of us know the soaring excitement of a roller-coaster ride or of plunging into a lake from a high ledge. I wonder if flying is exhilarating to birds (and bats) and what it might feel like when they make their inevitable first launch from a nest (see page 179).

Biologists have suggested that thrill-seeking behavior may be adaptive for another reason—that it is attractive to the other sex. This may explain why men are more likely than women to go bungee jumping, mountain climbing, or white-water rafting. Risky feats that demonstrate courage may send the message *I am a good protector*. If they also require strength and athleticism, they may convey a human's version of the peacock's display of physical fitness.

It is not clear to me why gray squirrels take unnecessary risks. But they do, and they seem to enjoy it. The woods behind my house are home to many squirrels. I often see them leaping between the trunks of trees spaced five or more feet apart, even though it would be only a short detour to go via the ground. They will also nose up to the window or screen behind which crouch my two cats, riveted on the cheeky rodents. I've not tried to ascertain whether male squirrels are more prone to this behavior than females, though evolutionary theory predicts that they should be. On one occasion, one of the cats jumped onto the screen, sending a sassy squirrel skittering away. But the rodent was back at the door two minutes later. Was he trying to impress others and improve his social prestige or perhaps seeking an adrenaline high?

Other animals play dangerous games, too. Orangutans perform a rather risky maneuver called snag-riding.[8] When in a tree, they deliberately reach out for a dead or weakened sapling and, if it falls, ride it to a neighboring branch. I've watched gulls, crows, and jackdaws tearing earthward from great heights before swooping up at the last moment.[9] There is no obvious survival function to this behavior, which leaves me wondering if they do it simply for the thrill of speed, as a human skydiver might.

Substance abuse is another thrill-seeking behavior, one which many humans and some animals participate in, despite there being no clear adaptive benefit. I have no photos of drunk elephants or inebriated birds in this book (intoxication is hard to detect in a still image), but there are widespread stories of animals getting drunk on the alcohol from rotting fruits. This is certainly not beneficial for birds, although there are numerous anecdotal accounts of inebriation in many frugivorous species. Intoxicated birds have difficulty perching and flying and are slower to notice and react to danger. Cedar waxwings (see page 58) have been known to die from alcohol poisoning after gorging on fermented berries.

It is likely that birds get intoxicated accidentally. The same may not be true of elephants, who have a reputation for getting drunk on fermenting marula fruits. Once again, however, the evidence does not come from rigorous scientific studies, and a team of scientists from the University of Bristol has questioned the notion that marulas could be making elephants drunk, claiming that the pachyderms would have to eat four times their usual meal size to be affected.[10] However, the researchers don't deny elephants' occasional tipsiness and suggest that the source could be a toxin found in beetle pupae under the marula tree bark, which elephants also eat.

Lemurs and capuchin monkeys are known to intoxicate themselves by grabbing and mouthing millipedes.[11] In response to the assault, the millipedes release potent defensive chemicals, which the primates rub against their fur. These are thought to act as effective repellants against parasites. But some practitioners take the behavior a step further; their mouthing and rubbing become more frantic until the chemicals send them into a euphoric-looking stupor. Then they become droopy and drool profusely and, according to some reports, may even pass the millipede from one to another before discarding it, usually not mortally wounded.[12]

I hesitated to include a section on curiosity in this book. There's no certain connection between this emotion and feeling pleasure, though I can't help thinking that an elk's first taste of snow and a monkey's fascination with water bring a certain form of pleasure (see pages 181 and 183, respectively). At the very least, curiosity involves awareness—a conscious engagement with one's surroundings. From a hotel balcony in northern India I watched a male parakeet sidle up to a female on a wire, where he courted her before they mated unhurriedly for forty-five seconds. While this was going on, two males on the wire just above them watched intently, their heads and bodies leaning toward the amorous couple.

Curiosity is no exception to most of the other pleasurable phenomena in this book in having an adaptive basis. Undoubtedly, it may be dangerous to explore something mysterious or partially unknown, especially if that something is another creature. But curiosity is an important aid to learning, particularly for younger individuals—which is probably why they tend to exhibit more curiosity about things than do adults. As a graduate student I studied eavesdropping by bats, who listen in on other bats' echolocation calls to find good patches of flying insects to feed on. It appears that young bats are much more responsive to these signals, perhaps because they have more to learn about the choice foraging locations that adults already know.[13] Fishes, it seems, practice a similar art. Juveniles tune in to the crackling sounds that emanate from reefs. This cueing allows them to find reef habitats from the open ocean. The snapping of shrimp claws and other sounds distinctive to reefs can be heard up to twenty kilometers away. Scientists discovered this skill by setting up artificial reefs with underwater speakers and broadcasting sounds recorded from real reefs.[14]

Unlike curiosity, humor is unambiguously pleasurable. The question is: are we the only ones who experience it? The most striking examples of a sense of humor among animals come from great apes who have been taught to communicate with sign language. In *Pleasurable Kingdom*, I provide several examples. I'll share just one here:

> When asked to identify the color of a white towel held up by a teacher, Koko [the gorilla matriarch of the Gorilla Foundation] signed "red." When asked again, she repeated this three times. Then, grinning, she plucked off a bit of red lint clinging to the towel, held it up to the trainer's face and signed "red" again.

Chimpanzees laugh in situations that tend to evoke human laughter, such as when playing with one another or with humans or when tickled.[15] All of the great apes—chimps, bonobos, gorillas, orangutans, and humans—share the tickle spot under the arms, and infant apes respond to a mother's tickling with a play face and laughter.[16]

Two American neuroscientists formerly at Bowling Green State University, Jaak Panksepp and Jeffrey Burgdorf, have presented evidence from both behavior and brain activity to support their hypothesis that rats use a form of laughter to signal to other rats that they desire to play. Rats trained to expect a tickle ran to the researcher's hand, on average, four times as quickly as did rats trained to expect a stroke on the neck, and they made seven times more playful ultrasonic chirps.[17]

A special pant made by domestic dogs is also believed to be a form of laughter; when Patricia Simonet and colleagues at the University of Nevada, Reno, broadcast recordings of it in an animal shelter, other dogs became playful and more sociable and exhibited lower stress levels.[18] Who knows what other forms of animal humor—or other sources of pleasure for animals with different sensory worlds—might be out there awaiting our discovery?

WEEVIL (Curculionidae sp.), Tatry Mountains, Poland. A tiny yellow weevil climbs a dew-dropped jungle of grass blades—a jewel among jewels. It is a reminder that some of the loveliest things come in small sizes. This image may or may not celebrate a beetle's feelings, but it undoubtedly resonates with our own. Photo: Andrey Antov.

ORANGE OAKLEAF BUTTERFLY (*Kallima inachus*), Magic Wings Butterfly Conservatory, South Deerfield, Massachusetts. Scientists debate whether an insect can feel anything, let alone any sort of pleasure on encountering a nectar-filled flower. What is not in debate is that we can delight in the esthetic beauty of both the butterfly and the hibiscus. Photo: Andrey Antov.

DOMESTIC CALF (*Bos taurus*), Farm Sanctuary, Orland, California. Before his rescue, Whitaker was most likely on his way either to slaughter or to a veal farm. He was soon nursed back to health at a California sanctuary for formerly farmed animals. Here, on the first day of his new life, Whitaker enjoys an open-air romp at his new home. (Another photo of Whitaker appears on page 12.) Photo: Connie Pugh.

ASIAN BLACK BEAR ("Moon bear"; *Ursus thibetanus*), Animals Asia Foundation's Moon Bear Rescue Centre, Chengdu, Sichuan, China. When the top photo was taken in 2000, the bear, eventually named Jasper by his rescuers, had been confined to this "crush cage" for fifteen years. Bile was drained from his gall bladder via a surgically implanted catheter, then sold as a medicine. The woman, a British national named Jill Robinson and founder of the Animals Asia Foundation, has since been honored by Queen Elizabeth for her efforts to negotiate the release of some of these bears. We may rightly be appalled at such treatment of a sentient animal, yet conditions are comparable for the equally sentient animals kept in factory farms and sold for meat. The lower photo is Jasper eight years later in his new home. Photos © Animals Asia Foundation. Bottom photo: Rainbow Zhu Ke.

KEA (*Nestor notabilis*), Parapara Peak, Golden Bay, New Zealand. Keas are legendary for curiosity and mischievousness. This photo was taken about four thousand feet above sea level. On reaching the site, the photographer was followed along a ridge by a group of about fifteen keas. They eyed him as he sat enjoying the view. This particular bird had more courage than the others and decided to taste his boot. Keas have a well-earned reputation for stripping the rubber linings from parked vehicles. Photo: Ben King.

OSPREY (*Pandion haliaetus*), Columbia River, Washington. As ground-bound mammals, we might find the idea of launching off a high aerie both thrilling and terrifying. For this recently fledged osprey taking off on her first flight, I surmise that the feelings are similar. As a bird genetically programmed for flight, however, she may have a powerful urge to lift off on wings aching for use, following weeks of practice flapping in the nest. And as her confidence blooms, we can imagine that the exhilaration of being airborne soon overtakes the fear. Photo: Ken Archer.

SWIFT FOX (*Vulpes velox*), Buffalo Gap National Grasslands, South Dakota. I did not choose this photo because it expresses pleasure. Indeed, how are we to know what this fox is feeling as he bounds across a field? I chose it because it expresses a fundamental value: freedom. Photo: Thomas D. Mangelsen.

SPRINGBOK (*Antidorcas marsupialis*), Etosha National Park, Namibia. A pair of male springboks exhibit a behavior known as stotting or pronking: performing repeated high, stiff-legged jumps by taking off on all fours at the same time. It gives the impression of a bouncing ball. Theories about its function include getting a better vantage point from which to spot predators, startling them, or simply letting them know *you've been seen and I'm too fit to bother chasing*. In support of the last theory, cheetahs have been shown to abandon hunts more often when the quarry stots, and the chances of any predator making a kill are less when stotting occurs. Predator contexts don't suggest that stotting is pleasurable, but it is also observed in social contexts, including play. Photo: Nate Chappell.

ELK (*Cervus canadensis*). A young bull elk engages in an act of playful curiosity commonly performed by young children—sticking out a tongue to catch snowflakes. Photo: Mark Peters.

BOTTLENOSE DOLPHIN (*Tursiops truncatus*), off the coast of Long Beach, California. These bottlenose dolphins were wake surfing, an aquatic version of the drafting that bicyclists and race-car drivers perform on land. They followed the boat for several minutes, matching its speed of about fifteen to twenty knots. An adaptationist interpretation is that they are saving energy. An experiential interpretation is that they are having a blast. Photo: Mike Moss.

BARBARY MACAQUE (*Macaca sylvanus*), Gibraltar, Spain. Some macaques show an intense fascination with water—its appearance, its movements, and its feel. The attention of this Barbary macaque was held completely for several minutes as she repeatedly splashed, apparently enchanted with the feel of the water and the consequence of the action. Photo: Andrew Forsyth.

DOMESTIC DOG (*Canis familiaris*). A large dog exhibits the joyous exuberance of a child (or a dog!) following a deep snowfall. Photo: David Buck.

CONCLUSION: IMPLICATIONS OF ANIMAL PLEASURE

When I set out to write this book, I considered whether it should be just a feel-good book or whether it should say something more. I soon decided on the latter. There is nothing wrong with feeling good for its own sake, and I hope people derive much pleasure and joy from this book. But, as the moon bear on page 177 reminds us, all is not well in the human relationship with animals. Change is needed.

This closing chapter is less focused on pleasure as such than on its significance to our appreciation and understanding of the animal kingdom. Broadly speaking, humankind's relationships with animals are profoundly out of step with what we now know of their awareness, perceptiveness, and sensitivity. Awareness of their pleasure enriches our view of them and shows us we are that much more similar to animals—from cats and dogs to pigs and chickens—than we once believed.

OUR TROUBLED RELATIONSHIP WITH ANIMALS

Humankind wields enormous power over animals, and for a long time now we've been abusing that privilege. Throughout history we have excluded them from our circle of moral concern. This is convenient, for it gives us free reign to eat them, wear them, and use them in harmful experiments, for entertainment, and for profit. It is a bleak legacy. Each year we kill more animals than in the previous year. Currently, the toll is approaching sixty billion land animals and a comparable number of fishes.[1]

Most of the animals we kill don't need to die—in particular, about 98 percent are killed to be eaten, and most of us can choose to eat other foods. There are also other ways to clothe ourselves, to advance science, to gain our own pleasures. Getting one's pleasure at the expense of another's is, to me, a poor equation for life. We can choose to live our lives in myriad ways that will help animals, or at least not harm them.

The roots of our ill treatment of animals lie in a number of prejudices. Long is the list of reasons used to justify our dominion over them: our intelligence, our language, our spirituality, our technology, our tool use, and our culture, to name a few. It is these same sorts of prejudices that justified colonialism, fostered slavery, and barred women from the right to vote. But the real arbiter of whether or not a being deserves respect and compassion is sentience—the capacity to feel. Being sensate to pleasures and especially to pains is the true currency of ethics.

Our particular intelligence has also given us an advanced capacity for moral thinking. Gradually, we are coming to question practices that make other feeling beings suffer for our interests. Neither might nor bright makes right. Otherwise, we must concede that if a more intelligent race arrived from outer space, they would have the right to torture and kill us if they so pleased. Hopefully, it would not please them to do this, but if it did, few of us would deem such a thing fair and just.

We've tended to define intelligence in terms that place humans at the top. No doubt, humans excel at technology, art, and literature, for example. But our smarts have also led to folly, such as the creation of weapons of mass destruction and the extermination of species. If intelligence were defined as the pursuit of happiness, perhaps dolphins or crows might outsmart us. Then there is perceptual intelligence. Bats leave us in the dust in interpreting echoes, dogs smells, and sea turtles the physical world of the ocean. When we pay close attention with an open mind to what animals do, we discover more interesting aspects of their behavior and their mental abilities. Animals are as intelligent as they need to be, and they are good at doing things important to their survival. Evolutionary events are mostly chance events. In the evolutionary sweepstakes, we got lucky. We can control dolphins because we happened to evolve hands and they didn't.

A CHANGING CLIMATE

There is another kind of respect we can show animals: that which comes from their being part of that broader reality we refer to as nature. Humankind's growing concern for protecting the vitality of natural systems is inevitable, in hindsight. We are animals in the biological sense, and we are no less dependent than other creatures on healthy, functioning ecosystems. As our numbers swell, we impinge more on natural spaces. And because our industrial activities are often not harmonious with nature, disruptions follow. Perhaps the most significant of these—climate change—has emerged only in the past century or so, and it is only in recent years that humankind has acknowledged global warming as real and serious.

Beyond the relatively obvious but vague notion that treating nature poorly is bad for the environment and for us, there is a concrete connection between our treatment of animals and the problem of climate change. In its groundbreaking report of November 2007, the Intergovernmental Panel on Climate Change (IPCC) concluded—based on an extensive, six-year-long analysis of data—that the contribution to greenhouse gas emissions from animal agriculture (about 18 percent) exceeds that of the entire global transportation sector (13.5 percent). In other words, what we purchase at the supermarket impacts the environment more than the means of transportation we choose to get there. Burgers and bacon are warming the planet even more than cars and airplanes.

Why is animal agriculture such a problem? The reason is that it is resource-costly to feed the human masses on animal protein. Energy flow through food chains is surprisingly inefficient: only about 2 percent of the energy in one level is transferred into the biomass at the next level.[2] It takes many tons of grass to sustain a ton of grazers and many tons of grazers to feed a ton of meat eaters. This is why carnivorous lions are so much rarer than herbivorous wildebeests and geese much more common than eagles.

The IPCC report was not prepared by vegetarians. The data were compiled, analyzed, and assessed by a team of international experts from a variety of fields, organized jointly by the World Meteorological Organization and the United Nations Environment Programme. Along with former U.S. vice president Al Gore, whose work has done so much to raise the visibility of climate change, the IPCC was awarded the 2008 Nobel Peace Prize.

But few of us are familiar with the IPCC and fewer still are aware of the meat connection to global warming. Because reducing or eliminating meat consumption is not yet an attractive idea for most omnivores, the message isn't being touted. People like the taste of meat, and they are in no hurry to give it up. (Note the role of pleasure in helping to drive climate change!) I attended a lecture by one of the IPCC authors. He spoke for eighty minutes to an audience of seven hundred high school science teachers, effectively outlining the compelling evidence for climate change and finishing up with some suggestions for personal action, including living closer to work, taking public transport, and using a bicycle for local errands. He made no mention of animal agriculture or meat consumption.

Denial doesn't make a problem go away. Knowledge is power, and if we are to turn the tide toward a healthier, more livable planet for all species (humans included), then we must acknowledge the role of the consumption of animal products and adopt alternatives.

A CHANGING CULTURE

Our power to change is one of our most noble and promising traits, and cultural change happens much more quickly than evolutionary change. The campaign to end institutionalized slavery was long and hard, but it was won in the course of about one hundred years, an eyeblink in evolutionary time. Thanks to the speed of cultural change, we've seen women's rights enshrined in the legislation of most of the world's countries in the one hundred fifty years since the dawn of the suffragist movement. And once it got rolling, the American civil rights movement overturned legalized racism in less than two decades. These efforts began with a small, dedicated group of people, then swelled under the power and correctness of their moral stance. The struggle to grant sentient animals freedom from factory farms, leghold traps, laboratory cages, and abattoirs is far more daunting for several reasons: the scale, scope, and power of the industries that rely on exploiting animals; the deeply ingrained idea that language, technology, and power are the only legitimate bases for thinking a species is morally relevant; and the fact that the victims are unable to articulate their plight. All of these factors weigh strongly against any quick, decisive change for animals.

The momentum, however, is beginning to turn in the animals' favor. The Humane Society of the United States reported a record 121 new state-level animal protection laws enacted in 2009, surpassing the previous record of 91 new laws enacted in 2008.[3] These laws are generally quite modest; they do not (yet) challenge animals' legal designation as human property. They involve such things as providing animals in factory farms enough space so that they can stand up and turn around (California), requiring the proper labeling of garments containing animal fur (Delaware), making dogfighting a felony-level crime (Georgia), and limiting to seventy-five the number of adult dogs allowed in a puppy mill (Louisiana). The European Union is implementing more far-reaching changes, the most significant of which is a phaseout of factory farm practices by the year 2012.

TAKING THEIR PERSPECTIVE

Empathy may be our most important emotion along the path to a more compassionate relationship with animals. To relate to others' feelings is to consider their experience, to try to imagine things from their perspective. For me, the gray seal below peering out from between two breaking waves conveys the essence of another's perspective. When I peer into those large, liquid, black eyes I wonder what the view looks like to her. Frozen in the moment Arthur Sevestre took the photo, I, the seal,

ATLANTIC GRAY SEAL (*Halichoerus grypus*), Donna Nook, United Kingdom. Our human experience provides only a limited perspective on a wild seal's experience of life. But our advanced intellect and enormous capacity to empathize and reflect allow us to imagine what it might be like. This image, for me, illustrates that a seal—as an autonomous being with senses and feelings—has her own perspective. Photo: Arthur Sevestre.

am gazing through the briefly parted curtain at the biped holding the camera on the shore. I hear the water's gushings and burblings. I smell and taste the brine. I feel the sluice of the cascading water against my fur, but I am cozily insulated against the cold by layers of enveloping blubber.

These fancies are informed by my own sensory experiences. But this exercise in sensory perspective taking has limitations. I don't possess a seal's brain. I don't know what space and time feel like to her, though I suspect they feel similar to how they feel to me. I don't know how much she dwells on her past or considers the future. Scientific studies have shown that seals are highly intelligent and have strong memories.[4] Is she feeling curious? Is she looking forward to lying out in the sun this evening on a rocky outcrop among others of her kind—individuals whom she knows? Perhaps her emotions are preoccupied by a sweeping current of love for a fellow seal she has been courting with, like the one on page 113.

Animals also have sensory experiences tuned to channels beyond our bandwidth. Those catlike whiskers around our seal's muzzle allow her to detect the turbulence trails left by fishes who may have swum past minutes before—a useful adaptation for foraging in deep waters where light barely penetrates and visibility is poor or nonexistent. We know the feeling of water movement from the sensation of hairs on our bodies responding to the turbulence of a current or a fellow swimmer. But I struggle to imagine the seal's experience of tracking the fish's movements and perhaps smelling or tasting its presence.

Because this book is meant for humans to enjoy, its message is part of its function. Ken Archer's photograph of a female cutthroat trout on page 190 thrusting her way up a shallow stream brings me immense pleasure. From the moment I set eyes on it, I knew it must appear in this book. This photo is a visual poem. It speaks to each of us uniquely. For me, it speaks of energy, wildness, and striving. The arch of her sides, the glassy ripple of water cresting her back, and the blurred, pale blue bubbles in the foreground convey her power against one of nature's inexorable forces. The roiling gold toward her tail invokes a burst of gasoline-fed flame fueling her passage. The glittering rows of tiny scales on her sides, the crimson, marbled sheen of her crescent gill cover, and the sprinkling of bold black dots—to my senses these are expressions of nature's passion.

If you look closely you can also see the faint lateral line cutting across the trout's scales at the midline. Collectively, these specialized scales constitute an organ for sensing movement and vibrations in the surrounding water. You may also have noticed that there is another trout immediately beneath her, its dorsal fin emerging from the water and the telltale speckles glinting below the surface. These fishes are working hard. They might not be having fun, but they are highly motivated, and for some there is the glow of sex at the end of their journey, when they will pair and spawn.

We underestimate fishes. They began evolving long before mammals, so they are considered primitive and unworthy of our moral consideration. But fishes did not stop evolving when their descendants moved onto land. Today they represent a magnificent diversity of species, outnumbering all other vertebrate groups combined.

Because fishes don't make facial expressions, because they don't scream or shout, many people continue to deny that they are capable of pain or suffering. But fishes manifest their fear and pain in

other ways, including the release of fear and pain chemicals.[5] Fishes have long-term memories, they recognize familiar individuals and have social preferences, they cooperate, and they have disputes and then reconcile.[6] Rapidly mounting scientific evidence shows them to be sentient like other vertebrates.[7]

It is my hope that readers will be more willing and able to connect emotionally with animals by appreciating the pleasurable aspects of their lives. Sadly, most human beings today, particularly in

CUTTHROAT TROUT (*Salmo clarki*), Lamar Valley, Yellowstone National Park, Wyoming. Cutthroat trout surge upstream to spawn. Can they anticipate where they are going and why? The conventional thinking is that they cannot, but new scientific studies show that fishes think and feel. We have hastened to deny this, although prudent, ethical thinking warrants a more open-minded view. Photo: Ken Archer.

urban settings, don't get close to any chickens, pigs, or salmon unless they are at the end of a fork. I am fortunate: the sanctuary near my home (see the chapter on touch) affords me a regular opportunity to connect with animals usually only encountered as anonymous, shrink-wrapped hunks of muscle at the supermarket. Despite being a professional ethologist, I rarely leave that sanctuary without having learned something new about these animals and their experiences. It was there that I first witnessed a rooster's *come hither* call, which brings a nearby hen running for a free handout. While helping medicate the pigs there I noticed that those without arthritis didn't look up expectantly from their slumber—because they know it isn't part of their routine to receive a third of a banana stuffed with anti-inflammatory pills. Only the pigs who are regularly dosed anticipated their treats. And it was there on cold winter mornings that I observed the magnetic effect the underside of a heated water bowl has on wild mice—as many as ten of which I've watched scatter across the hay when I lifted the bowl to clean and refill it.

Poplar Spring Animal Sanctuary's mission is to teach respect for animals as fellow autonomous beings with a will to live and a lust for life. No traps are laid out for the rodents who sometimes gravitate to the barns and sheds for the spillings left by the residents. Instead, hole-punched metal strips are hooked to the rims of water buckets so a thirsty mouse who gets an unintended dunking has an escape route from drowning. When I caught a black rat snake in the henhouse, the policy was to release it either back into the woods or in one of the barns to discourage the rodents from getting too comfortable. And no fork meets meat there. In keeping with its ethic of respect for all animals, the place is vegan.

LIVES WORTH LIVING

As pleasure seekers, animals have interests that go beyond the mere avoidance of pain; they desire a good quality of life. The implications of this fact are profound. Because animals feel good things, their lives are worth living. Pleasure gives their lives intrinsic value—that is, value to themselves beyond any utilitarian worth they may have for us.[8]

When we realize that other creatures value their freedom and that they have good days and bad days, we may be moved to treat them better. As I've argued in this chapter, we behave poorly when we deprive animals of the opportunity to enjoy their existence, particularly if we do so for relatively trivial reasons—which surely account for over 99 percent of our exploitation of animals. A pleasur-

HUMAN (*Homo sapiens*) and BULL (*Bos taurus*), Farm Sanctuary, Orland, California. The photographer's daughter shares the pleasure of contact with a young rescued bull named Harrison. He knows Simone, a forensic scientist and regular visitor to the sanctuary. I have also met Harrison, a gentle hedonist who loves to have his neck massaged. Photo: Connie Pugh.

able kingdom (or queendom, if you prefer) is a richer place than a neutral one; all animals—whether free or domesticated—deserve the opportunity to pursue their pleasures, and their capacity for feeling good warrants a more compassionate ethic from us.

All of the animals featured in this book are individuals. This is obvious, but it is worth mentioning because too often we view animals only as members of a species or a population. Species and populations are useful concepts, but they don't take into consideration animals' sentience. Species and populations don't feel pains or pleasures; only individuals do. So when we consider animals with regard to their capacity to feel, we must consider them as separate and unique. As surely as they each have a biology, each also has a biography.

A happy consequence of the individual-based perspective of animals, as opposed to the species-based one, is that every good deed is meaningful and significant. An experience recounted by the anthropologist Loren Eisley serves to illustrate. Eisley once encountered a young man on a beach strewn with millions of stranded starfish dying in the sun. As the young man tossed another starfish into the surf, Eisley politely informed him that there were miles and miles of beach and that his actions could not possibly make a difference. At this, the young man bent down, picked up yet another starfish, and, as it met the water, said: "It made a difference for that one."[9]

Animal pleasure is a good thing. That animals have positive experiences corrects the common misconception that nature is only a harsh and cruel place where avoiding predation and starvation defines an animal's daily existence. To be sure, life for all animals (humans included) is challenging, and for every individual who lives to reproduce, many more fall by the wayside. But life, at any age, has rewards.

The three American black bears shown on pages 194 and 195 encapsulate the idea of an exultant ark. The blissful contentment of a mother bear as she suckles her two cubs shows a wild animal in a peaceful moment of pleasure. The two cubs, gazing apprehensively over their shoulders to ensure that mother approves or perhaps that she is coming with them, convey the emotional tenor of their lives. Finally, as the pair turn away and set forth on another adventure, we are reminded of the irrepressible optimism that is life.

The eighteenth-century poet William Cowper understood the value of life for all who experience it. And so I close with these words from his 1785 poem *The Task:*

> The bounding fawn that darts across the glade
> When none pursues, through mere delight of heart,
> And spirits buoyant with excess of glee;
> The horse, as wanton and almost as fleet,
> That skims the spacious meadow at full speed,
> Then stops and snorts, and throwing high his heels
> Starts to the voluntary race again . . .

AFRICAN ELEPHANT (*Loxodonta africana*) and **CATTLE EGRET** (*Bubulcus ibis*), Amboseli National Park, Kenya. Four egrets hitch a ride on a bull elephant. None of these creatures owns anything. They are unfettered and completely free. Like us, they are faced with the vicissitudes of life. But it is life worth living. Photo: Karl Ammann.

BONOBO (*Pan paniscus*), Bonobo sanctuary near Kinshasa, Democratic Republic of the Congo. A bonobo reflects back at his closest living evolutionary relative. Photo: Karl Ammann.

BLACK BEAR (*Ursus americanus*), Wawayanda State Park, New Jersey. These three images embody the broader message of this book. A mother bear performs the primal nurturant act of all mother mammals: nursing her babies. Sated, the youngsters look back to make sure that their caregiver is still with them. Then they set off to explore the unknown world. Photos: Susan Kehoe.

NOTES

INTRODUCTION

1. Struhsaker 1967.
2. Brown et al. 2006.
3. Galef and Whiskin 2003.
4. Burghardt 1991.
5. Bekoff 2006.
6. De Waal 2001: 69.
7. Richner 1989.
8. Cabanac 1971.
9. Dehnhardt et al. 2001.
10. Heyers et al. 2007.
11. Panksepp 1998.
12. Silverman 2009.
13. Persinger 2003.
14. Danbury et al. 2000.
15. Barr et al. 2008.
16. On crabs: Patterson et al. 2007; on octopuses: Sinn et al. 2001.
17. See Panksepp and Huber 2004; Nathaniel et al. 2009.
18. Balaskó and Cabanac 1998.
19. Cabanac 1971.
20. Cohen and Tokieda 1972.
21. Burgdorf and Panksepp 2007.
22. Rollin 1989: 144.
23. Uexküll 1934.
24. On reptiles and amphibians: Cabanac 2005; on fishes: Tebbich et al. 2002; on chimpanzees: Inoue and Matsuzawa 2007.
25. See Balcombe 2010.
26. See Dawkins 1998; Olsson and Dahlborn 2002.
27. Neuringer 1969.
28. Balcombe 2006.
29. Sapolsky 2001.
30. Moss 1983.
31. Evans and Evans 1999.
32. Gyger and Marler 1988.
33. Kendrick et al. 2001.
34. Da Costa et al. 2004.
35. Tate et al. 2006.
36. Kohler and Wehner 2005.
37. See Mather and Anderson 1993; Mather 1992; Anderson et al. 2010; Kuba et al. 2003.
38. Linden 2003.
39. Kuba et al. 2003.
40. Mather 1992.
41. Brown et al. 2006.
42. Dugatkin and Wilson 1992.
43. Brown et al. 2006.
44. Teyke 1985, 1989.
45. Largiader et al. 2001.
46. Doutrelant et al. 2001.
47. Aronson 1972.
48. Rose 2007.
49. Sneddon 2009.
50. Sneddon 2003.
51. Frisch 1938; Brown and Chivers 2006.

PLAY

1. Burghardt 1984; Caro 1988.
2. On reptiles and fishes: Ortega and Bekoff 1987; Burghardt 2005. On primates: Burghardt 2005.
3. Pozis-Francois et al. 2004.
4. Murdochs 2008; Skelhorn 2008.

5. Brown 1994.
6. Bekoff and Byers 1998.
7. Panksepp and Burgdorf 2003.
8. On rats: Pellis 2002; on wallabies: Watson and Croft 1996.
9. Flack et al. 2004.
10. Brosnan and de Waal 2003.
11. Range et al. 2008.
12. Balcombe 2010.
13. De Waal 2005.
14. On mole rats: Locantore 2002; on jackdaws: de Kort et al. 2006.

FOOD

1. Berridge 1996.
2. Cabanac 2005.
3. Laska et al. 2000.
4. Laska et al. 2003.
5. Laska and Hernandez Salazar 2004.
6. Laska 1994; Laska 1996.
7. Zuk 2003: 151.
8. Cabanac and Johnson 1983.
9. Balaskó and Cabanac 1998.
10. Johnston and Fenton 2001.
11. Fa and Cavalheiro 2008.
12. Lehrer 2007.
13. Strieck 1924.
14. Yasuoka and Abe 2009.
15. Mitani and Watts 2001.

TOUCH

1. Winfrey 2008: 8.
2. Uhlenbroek 2002.
3. Feh and de Mazières 1993.
4. Burghardt 2004.
5. Tebbich et al. 2002.
6. Cheney et al. 2008.
7. Bshary and Shaeffer 2002.
8. Bshary and Wuerth 2001.
9. Bshary 2006.
10. Deeble 2003.

COURTSHIP AND SEX

1. De Waal and Lanting 1997.
2. Roughgarden 2004.
3. See Dixson 1998; Chevalier-Skolnikoff 1974.
4. Zuk 2003.

5. Winterbottom et al. 2001.
6. Bagemihl 1999.
7. Bagemihl 1999.
8. Winkler 2009.
9. Liley 1984.
10. Alfieri and Dugatkin 2006.

LOVE

1. Peterson 2004.
2. McFarland 1987; Bekoff 2004.
3. Van der Gucht and Hof 2007.
4. Coghlan 2006.
5. Hakeem et al. 2008.
6. Bradshaw et al. 2005.
7. Bradshaw 2009.
8. Douglas-Hamilton et al. 2006.
9. Lorenz 1978.
10. Goodall 2001.
11. Heinrich 1999: 341.
12. Behr and von Helversen 2004.
13. Huang and Hessler 2008.
14. McFarland 1987.
15. On sexual arousal in marmosets: Ferris et al. 2004; on romantic love in women: Bartels and Zeki 2000.

COMFORT

1. Cabanac 1971.
2. E.g., Carrascal et al. 2001.
3. Heinrich 2003.
4. Panksepp and Burgdorf 1999.
5. Montgomery 2006.
6. Montgomery 2009.

COMPANIONSHIP

1. On boars: Quenette and Gerard 1992; on kangaroos: Pays et al. 2009; on ostriches: Bertram 1980.
2. Slobodchikoff et al. 1991.
3. Frederiksen and Slobodchikoff 2007.
4. Millett and Pratt 2005.
5. Moss 1983.
6. Wilkinson 1992.
7. On cows recognizing individuals: Coulon et al. 2009; on cows having favorites: Young 2003.
8. Griffiths and Ward 2006.
9. Engh et al. 2006.

10. Cheney and Seyfarth 2007.
11. Preston and de Waal 2002.
12. De Waal and Lanting 1997.
13. Langford et al. 2006.

OTHER PLEASURES
1. Pollan 2001: 19.
2. Anderson et al. 2005.
3. Ghirlanda et al. 2004.
4. Thornhill and Gangestad 1999.
5. Weatherhead and Robertson 1979.
6. On bird cognition, see, e.g., Clayton and Dickinson 1998; Pepperberg 2000. On bird emotions, see, e.g., Heinrich 1999; Burger 2001.
7. E.g., Bateson and Matheson 2007; Matheson et al. 2008.
8. Van Schaik et al. 2003.
9. Balcombe 2006.
10. Morris et al. 2006.
11. Birkinshaw 1999.

12. Downer 2002.
13. Balcombe and Fenton 1988.
14. Tolimieri et al. 2004.
15. Cardoso 2001.
16. Goodall 1986.
17. Burgdorf and Panksepp 2001.
18. Allan 2006.

CONCLUSION
1. Steinfeld et al. 2006.
2. Colinveaux 1979.
3. Humane Society of the United States 2010.
4. On sea lion intelligence: Genty and Roeder 2006; on sea lion memory: Schusterman et al. 1993.
5. Chivers and Smith 1998.
6. Brown et al. 2006.
7. Braithwaite 2010.
8. Balcombe 2009.
9. Eisley 1972.

REFERENCES

Alfieri MS, Dugatkin LA. 2006. Cooperation and cognition in fishes. In *Fish Cognition and Behavior,* ed. Brown C, Laland K, Krause J, pp. 203–222. Oxford, U.K.: Blackwell Publishing.

Allan C. 2006. Quieter shelter dogs? It's a laughing matter. *Animal Sheltering,* May/June, p. 6.

Anderson JR, Leighty KA, Kuwahata H, Kuroshima H, Fujita K. 2005. Are monkeys aesthetists? Rensch (1957) revisited. *Journal of Experimental Psychology: Animal Behavior Processes* 31: 71–78.

Anderson RC, Mather JA, Monette MQ, Zimsen SRM. 2010. Octopuses (*Enteroctopus dofleini*) recognize individual humans. *Journal of Applied Animal Welfare Science* 13: 261–272.

Aronson LR. 1972. Further studies on orientation and jumping behavior in the gobiid fish, *Bathygobius soporator. Annals of the New York Academy of Sciences* 188: 378–392.

Bagemihl B. 1999. *Biological Exuberance: Animal Homosexuality and Natural Diversity.* London: Profile Books, Ltd.

Balaskó M, Cabanac M. 1998. Behavior of juvenile lizards (*Iguana iguana*) in a conflict between temperature regulation and palatable food. *Brain, Behavior and Evolution* 52: 257–262.

Balcombe JP. 2006. *Pleasurable Kingdom: Animals and the Nature of Feeling Good.* London: Macmillan.

———. 2009. Animal pleasure and its moral significance. *Applied Animal Behaviour Science* 118: 208–216.

———. 2010. *Second Nature: The Inner Lives of Animals.* New York: Palgrave.

Balcombe JP, Fenton MB. 1988. Eavesdropping by bats: The influence of echolocation call design and foraging strategy. *Ethology* 79: 158–166.

Barr S, Laming PR, Dick JTA, Elwood RW. 2008. Nociception or pain in a decapod crustacean? *Animal Behaviour* 75: 745–751.

Bartels A, Zeki S. 2000. The neural basis of romantic love. *Neuroreport* 27: 3829–3834.

Bateson M, Matheson SM. 2007. Performance on a categorization task suggests that removal of environmental enrichment induces "pessimism" in captive European starling (*Sturnus vulgaris*). *Animal Welfare* 16S: 33–36.

Behr O, Helversen O von. 2004. Bat serenades—complex courtship songs of the sac-winged bat (*Saccopteryx bilineata*). *Behavioral Ecology and Sociobiology* 56: 106–115.

Bekoff M. 2004. *Encyclopedia of Animal Behavior.* Westport, Conn.: Greenwood Press.

———. 2006. *Animal Passions and Beastly Virtues: Reflections on Redecorating Nature.* Philadelphia: Temple University Press.

Bekoff M, Byers J (eds.). 1998. *Animal Play: Evolutionary, Comparative, and Ecological Perspectives.* New York: Cambridge University Press.

Berridge K. 1996. Food reward: Brain substrates of wanting and liking. *Neuroscience and Biobehavioral Reviews* 20: 1–25.

Bertram BCR. 1980. Vigilance and group size in ostriches. *Animal Behaviour* 28: 278–286.

Birkinshaw CR. 1999. Use of millipedes by black lemurs to anoint their bodies. *Folia Primatologica* 70: 170–171.

Bradshaw G. 2009. *Elephants on the Edge: What Animals Teach Us about Humanity.* New Haven, Conn.: Yale University Press.

Bradshaw GA, Shore AN, Brown JL, Poole JH, Moss CJ. 2005. Elephant breakdown. *Nature* 433: 807.

Braithwaite V. 2010. *Do Fish Feel Pain?* Oxford: Oxford University Press.

Brosnan SF, de Waal FBM. 2003. Monkeys reject unequal pay. *Nature* 425: 297–299.

Brown C, Laland K, Drause J. 2006. *Fish Cognition and Behavior.* Oxford, U.K.: Blackwell Publishing.

Brown GE, Chivers DP. 2006. Learning about danger: Chemical alarm cues and the assessment of predation risk by fishes. In *Fish Cognition and Behavior,* ed. Brown C, Laland K, Krause J, pp. 49–69. Oxford, U.K.: Blackwell Publishing.

Brown SL. 1994. Animals at play. *National Geographic,* December, p. 30.

Bshary R. 2006. Machiavellian intelligence in fishes. In *Fish Cognition and Behavior,* ed. Brown C, Laland K, Krause J, pp. 223–242. Oxford, U.K.: Blackwell Publishing.

Bshary R, Shaeffer D. 2002. Choosy reef fish select cleaner fish that provide high-quality service. *Animal Behaviour* 63: 557–564.

Bshary R, Wuerth M. 2001. Cleaner fish *Labroides dimidiatus* manipulate client reef fish by providing tactile stimulation. *Proceedings of the Royal Society of London B: Biological Sciences* 268: 1495–1501.

Burgdorf J, Panksepp J. 2001. Tickling induces reward in adolescent rats. *Physiology and Behavior* 72: 167–173.

———. 2007. The neurobiology of positive emotions. *Neuroscience and Biobehavioral Reviews* 30: 173–187.

Burger J. 2001. *The Parrot Who Owns Me: The Story of a Relationship.* New York: Villard.

Burghardt GM. 1984. On the origins of play. In *Play in Animals and Humans,* ed. Smith PK, pp. 5–41. Oxford, U.K.: Basil Blackwell.

———. 1991. Cognitive ethology and critical anthropomorphism: A snake with two heads and hognose snakes that play dead. In *Cognitive Ethology: The Minds of Other Animals,* ed. Ristau CA, pp. 53–90. San Francisco: Erlbaum.

———. 2004. Iguana research: Looking back and looking ahead. In *Iguanas: Biology and Conservation,* ed. Alberts AD, Carter RL, Hayes WK, Martins EP, pp. 1–12. Berkeley: University of California Press.

———. 2005. *The Genesis of Animal Play: Testing the Limits.* Cambridge, Mass.: Bradford Books.

Cabanac M. 1971. Physiological role of pleasure. *Science* 173: 1103–1107.

———. 2005. The experience of pleasure in animals. In *Mental Health and Well-Being in Animals,* ed. McMillan FD, pp. 29–46. Ames: Iowa State University Press.

Cabanac M, Johnson KG. 1983. Analysis of a conflict between palatability and cold exposure in rats. *Physiology and Behavior* 31: 249–253.

Cardoso SH. 2001. Our ancient laughing brain. *Cerebrum* 2(4): 15–30.

Caro TM. 1988. Adaptive significance of play: Are we getting closer? *Trends in Ecology and Evolution* 3: 50–54.

Carrascal LM, Diaz JA, Huertas DL, Mozetich I. 2001. Behavioral thermoregulation by treecreepers: Trade-off between saving energy and reducing crypsis. *Ecology* 82: 1642–1654.

Cheney DL, Seyfarth RM. 2007. *Baboon Metaphysics: The Evolution of a Social Mind.* Chicago: University of Chicago Press.

Cheney KL, Bshary R, Grutter AS. 2008. Cleaner fish cause predators to reduce aggression toward bystanders at cleaning stations. *Behavioral Ecology* 19: 1063–1067.

Chevalier-Skolnikoff S. 1974. The ontogeny of communication in the stumptail macaque (*Macaca arctoides*). *Contributions to Primatology* 2: 1–166.

Chivers DP, Smith RJF. 1998. Chemical alarm signaling in aquatic predator-prey systems: A review and prospectus. *Écoscience* 5: 338–352.

Clayton NS, Dickinson A. 1998. Episodic-like memory during cache recovery by scrub jays. *Nature* 395: 272–274.

Coghlan A. 2006. Whales boast the brain cells that "make us human." *New Scientist,* November 27.

Cohen PS, Tokieda FK. 1972. Sucrose water preference reversal in the water deprived rat. *Journal of Comparative and Physiological Psychology* 79: 254–258.

Colinvaux PA. 1979. *Why Big, Fierce Animals Are Rare.* Princeton, N.J.: Princeton University Press.

Coulon M, Deputte BL, Heyman Y, Baudoin C. 2009. Individual recognition in domestic cattle (*Bos taurus*): Evidence from 2D-images of heads from different breeds. *PLoS ONE* 4: e4441.

da Costa AP, Leigh AE, Man MS, Kendrick KM. 2004. Face pictures reduce behavioural, autonomic, endocrine and neural indices of stress and fear in sheep. *Proceedings of the Royal Society of London* 271: 2077–2084.

Danbury TC, Weeks CA, Chambers JP, Waterman-Pearson AE, Kestin SC. 2000. Self-selection of the analgesic drug, carprofen, by lame broiler chickens. *Veterinary Record* 146: 307–311.

Dawkins MS. 1998. Evolution and animal welfare. *Quarterly Review of Biology* 73: 305–328.

Deeble M. 2003. Hippo heaven. *BBC Wildlife* 21: 42–49.

Dehnhardt G, Mauck B, Hanke W, Bleckmann H. 2001. Hydrodynamic trail-following in harbor seals (*Phoca vitulina*). *Science* 293: 29–31.

Dixson AF. 1998. *Primate Sexuality: Comparative Studies of the Prosimians, Monkeys, Apes, and Human Beings.* Oxford: Oxford University Press.

Douglas-Hamilton I, Bhalla S, Wittemyer G, Vollrath F. 2006. Behavioural reactions of elephants towards a dying and deceased matriarch. *Applied Animal Behaviour Science* 100: 87–102.

Doutrelant C, McGregor PK, Oliveira RF. 2001. The effect of an audience on intrasexual communication in male Siamese fighting fish, *Betta splendens. Behavioral Ecology* 12: 283–296.

Downer, J. 2002. *Weird Nature.* London: BBC.

Dugatkin LA, Wilson DS. 1992. The prerequisites for strategic behaviour in bluegill sunfish, *Lepomis macrochirus. Animal Behaviour* 44: 223–230.

Eisley L. 1972. *The Unexpected Universe.* New York: Mariner.

Engh AL, Beehner JC, Bergman TJ, Whitten PL, Hoffmeier RR, Seyfarth RM, et al. 2006. Behavioural and hormonal responses to predation in female chacma baboons (*Papio hamadryas ursinus*). *Proceedings of the Royal Society B* 273: 1587.

Evans CS, Evans L. 1999. Chicken food calls are functionally referential. *Animal Behaviour* 58: 307–319.

Fa JE, Cavalheiro ML. 2008. Individual variation in food consumption and food preferences in St Lucia parrots *Amazona versicolor* at Jersey Wildlife Preservation Trust. *International Zoo Yearbook* 36: 199–214.

Feh C, de Mazières J. 1993. Grooming at a preferred site reduces heart rate in horses. *Animal Behaviour* 46: 1191–1194.

Ferris CF, Snowdon CT, King JA, Sullivan JM, Ziegler TE, Olson DP, Schultz-Darken NJ, Tannenbaum PL, Ludwig R, Wu Z, Einspanier A, Vaughan JT, Duong TQ. 2004. Activation of neural pathways associated with sexual arousal in non-human primates. *Journal of Magnetic Resonance Imaging* 19: 168–175.

Flack JC, Jeanotte LA, de Waal FBM. 2004. Play signaling and the perception of social rules by juvenile chimpanzees (*Pan troglodytes*). *Journal of Comparative Psychology* 118: 149–159.

Frederiksen JK, Slobodchikoff CN. 2007. Referential specificity in the alarm calls of the black-tailed prairie dog. *Ethology Ecology and Evolution* 19: 87–99.

Frisch Kv. 1938. Zur Psychologie des Fische-Schwarmes. *Naturwissenschaften* 26: 601–606.

Galef BG Jr, Whiskin EE. 2003. Preference for novel flavors in adult Norway rats (*Rattus norvegicus*). *Journal of Comparative Psychology* 117: 96–100.

Genty E, Roeder JJ. 2006. Self-control: Why should sea lions, *Zalophus californianus,* perform better than primates? *Animal Behaviour* 72: 1241–1247.

Ghirlanda S, Jansson L, Enquist M. 2004. Chickens prefer beautiful humans. *Human Nature* 13: 383–389.

Goodall J. 1986. *The Chimpanzees of Gombe: Patterns of Behavior.* Cambridge, Mass.: Belknap Press.

———. 2001. *The Chimpanzees I Love: Saving Their World and Ours.* New York: Scholastic.

Griffin, DR. 1976. *The Question of Animal Awareness.* New York: Rockefeller University Press.

———. 1984. *Animal Thinking.* Cambridge, Mass.: Harvard University Press.

———. 1992. *Animal Minds.* Chicago: University of Chicago Press.

Griffiths SW, Ward A. 2006. Learned recognition of conspecifics. In *Fish Cognition and Behavior,* ed. Brown C, Laland K, Krause J, pp. 139–165. Oxford, U.K.: Blackwell Publishing.

Gucht PR Van der, Hof E. 2007. Structure of the cerebral cortex of the humpback whale, *Megaptera novaeangliae* (Cetacea, Mysticeti, Balaenopteridae). *Anatomical Record* 290: 1–31.

Gyger M, Marler P. 1988. Food calling in the domestic fowl (*Gallus gallus*): The role of external referents and deception. *Animal Behaviour* 36: 358–365.

Hakeem AY, Sherwood CC, Bonar CJ, Butti C, Hof PR, Allman JM. 2008. Von Economo neurons in the elephant brain. *Anatomical Record* 292: 242–248.

Heinrich B. 1999. *Mind of the Raven: Investigations and Adventures with Wolf-Birds.* New York: HarperCollins.

———. 2003. *Winter World: The Ingenuity of Animal Survival.* New York: Ecco.

Heyers D, Manns M, Luksch H, Güntürkün O, Mouritsen H. 2007. A visual pathway links brain structures active during magnetic compass orientation in migratory birds. *PLoS ONE* 2: e937.

Huang Y-C, Hessler NA. 2008. Social modulation during songbird courtship potentiates midbrain dopaminergic neurons. *PLoS ONE* 3, e3281.

Humane Society of the United States. 2010. *Rescue, Reform, Results: Annual Report 2009.* Washington, D.C.: The Humane Society of the United States.

Inoue S, Matsuzawa T. 2007. Working memory of numerals in chimpanzees. *Current Biology* 17: R1004–R1005.

Johnston DS, Fenton MB. 2001. Individual and population level variability in the diets of pallid bats (*Antrozous pallidus*). *Journal of Mammalogy* 82: 362–373.

Kendrick KM, da Costa AP, Leigh AE, Hinton MR, Peirce JW. 2001. Sheep don't forget a face. *Nature* 414: 165–166.

Kohler M, Wehner R. 2005. Idiosyncratic route-based memories in desert ants, *Melophorus bagoti:* How do they interact with path-integration vectors? *Neurobiology of Learning and Memory* 83: 1–12.

Kort SR de, Emery NJ, Clayton NS. 2006. Food sharing in jackdaws, *Corvus monedula:* What, why and with whom? *Animal Behaviour* 72: 297–304.

Kuba M, Meisel DV, Byrne RA, Griebel U, Mather JA. 2003. Looking at play in *Octopus vulgaris.* In Warnke K, Keupp H, Boletzky Sv (eds.), *Coleoid Cephalopods Through Time. Berliner Paläontologische Abhandlungen* 3: 163–169.

Langford, DJ, Crager SE, Shehzad Z, Smith SB, Sotocinal SG, Levenstadt JS, Chanda ML, Levitin DJ, Mogil JS. 2006. Social modulation of pain as evidence for empathy in mice. *Science* 312: 1967–70.

Largiader CR, Fries V, Bakker TCM. 2001. Genetic analysis of sneaking and egg-thievery in a natural population of the three-spined stickleback (*Gasterosteus aculeatus* L.). *Heredity* 48: 459–468.

Laska M. 1994. Taste difference thresholds for sucrose in squirrel monkeys (*Saimiri sciureus*). *Folia Primatologica* 63: 144–148.

———. 1996. Taste preference thresholds for food-associated sugars in the squirrel monkey (*Saimiri sciureus*). *Primates* 37: 91–95.

Laska M, Hernandez Salazar LT. 2004. Gustatory responsiveness to monosodium glutamate and sodium chloride in four species of nonhuman primates. *Journal of Experimental Zoology* 301A: 898–905.

Laska M, Hernandez Salazar LT, Rodriguez Luna E. 2000. Food preferences and nutrient composition in captive spider monkeys, *Ateles geoffroyi. International Journal of Primatology* 21: 671–683.

Laska M, Scheuber HP, Hernandez Salazar LT, Rodriguez Luna E. 2003. Sour-taste tolerance in four species of nonhuman primates. *Journal of Chemical Ecology* 29: 2637–2649.

Lehrer J. 2007. *Proust Was a Neuroscientist.* New York: Mariner.

Liley NR. 1984. Sex change in fish found common. *New York Times,* December 4.

Linden E. 2003. *The Octopus and the Orangutan: New Tales of Animal Intrigue, Intelligence, and Ingenuity.* London: Plume.

Locantore J. 2002. The naked truth about mole-rats. *Smithsonian Zoogoer,* May/June, pp. 24–28.

Lorenz K. 1978. *The Year of the Greylag Goose.* New York: Harcourt Brace Jovanovich.

Mather J. 1992. Underestimating the octopus. In *The Inevitable Bond: Examining Scientist-Animal Interactions,* ed. Davis H, Balfour D, pp. 240–249. New York: Cambridge University Press.

Mather JA, Anderson RC. 1993. Personalities of octopuses (*Octopus rubescens*). *Journal of Comparative Psychology* 107(3): 336–340.

Matheson SM, Asher L, Bateson M. 2008. Larger, enriched cages are associated with "optimistic" response biases in captive European starlings (*Sturnus vulgaris*). *Applied Animal Behaviour Science* 109: 374–383.

McFarland D. 1987. *The Oxford Companion to Animal Behavior.* New York: Oxford University Press.

Millett RP, Pratt JP. 2005. Prairie dog language? *Meridian Magazine,* May 19.

Mitani JC, Watts DP. 2001. Why do chimpanzees hunt and share meat? *Animal Behaviour* 61: 915–924.

Montgomery S. 2006. *The Good Good Pig.* New York: Ballantine Books.

———. 2009. Personal communication, February 16.

Morris S, Humphreys D, Reynolds D. 2006. Myth, marula, and elephant: An assessment of voluntary ethanol intoxication of the African elephant (*Loxodonta africana*) following feeding on the fruit of the marula tree (*Sclerocarya birrea*). *Physiological and Biochemical Zoology* 79(2): 363–369.

Moss C. 1983. *Portraits in the Wild: Animal Behavior in East Africa* (2nd ed.). Chicago: University of Chicago Press.

Murdochs S. 2008. The last word: Foxy tale. *New Scientist* 2677: 81.

Nathaniel TI, Panksepp J, Huber R. 2009. Drug-seeking behavior in an invertebrate system: Evidence of morphine-induced reward, extinction and reinstatement in crayfish. *Behavioural Brain Research* 197: 331–338.

Neuringer A. 1969. Animals respond for food in the presence of free food. *Science* 166: 399–401.

Olsson AS, Dahlborn K. 2002. Improving housing conditions for laboratory mice: A review of "environmental enrichment." *Laboratory Animals* 36: 243–270.

Ortega JC, Bekoff M. 1987. Avian play: Comparative evolutionary and developmental trends. *Auk* 104: 338–341.

Panksepp J. 1998. *Affective Neuroscience.* Oxford: Oxford University Press.

Panksepp J, Burgdorf J. 1999. Laughing rats? Playful tickling arouses high frequency ultrasonic chirping in young rodents. In *Toward a Science of Consciousness,* vol. 3, ed. Hameroff S, Chalmers D, Kazniak A, pp. 124–136. Cambridge, Mass.: MIT Press.

———. 2003. "Laughing" rats and the evolutionary antecedents of human joy? *Physiology and Behavior* 79: 533–47.

Panksepp JB, Huber R. 2004. Ethological analyses of crayfish behavior: A new invertebrate system for measuring the rewarding properties of psychostimulants. *Behavioural Brain Research* 153: 171–180.

Patterson L, Dick JTA, Elwood RW. 2007. Physiological stress responses in the edible crab *Cancer pagurus* to the fishery practice of de-clawing. *Marine Biology 152:* 265–272.

Pays O, Goulard M, Blomberg SP, Goldizen AW, Sirot E, Jarman PJ. 2009. The effect of social facilitation on vigilance in the eastern gray kangaroo, *Macropus giganteus.* *Behavioral Ecology* 20: 469–477.

Pellis SM. 2002. Keeping in touch: Play fighting and social knowledge. In *The Cognitive Animal,* ed. Bekoff M, Allen C, and Burghardt GM, pp. 421–427. Cambridge, Mass.: MIT Press.

Pepperberg IM. 2000. *The Alex Studies: Cognitive and Communicative Abilities of Grey Parrots.* Cambridge, Mass.: Harvard University Press.

Persinger MA. 2003. Rats' preferences for an analgesic compared to water: An alternative to "killing the rat so it does not suffer." *Perceptual and Motor Skills* 96: 674–680.

Peterson B. 2004. Caregiving: Attachment behaviors. In *Encyclopedia of Animal Behavior,* vol. 1, ed. Bekoff M, pp. 175–177. Westport, Conn.: Greenwood Press.

Pollan M. 2001. *The Botany of Desire: A Plant's-Eye View of the World.* New York: Random House.

Pozis-Francois O, Zahavi A, Zahavi A. 2004. Social play in Arabian babblers. *Behaviour* 141: 425–450.

Preston SD, de Waal FB. 2002. Empathy: Its ultimate and proximate bases. *Behavioral and Brain Sciences* 25: 1–20.

Quenette P-Y, Gerard J-F. 1992. From individual to collective vigilance in wild boar (*Sus scrofa*). *Canadian Journal of Zoology* 70: 1632–1635.

Range F, Horn L, Viranyi Z, Huber L. 2008. Absence of reward induced aversion to inequity in dogs. *Proceedings of the National Academy of Sciences* 106: 340–345.

Richner H. 1989. Phenotypic correlates of dominance in carrion crows and their effects on access to food. *Animal Behaviour* 38: 606–612.

Rollin B. 1989. *The Unheeded Cry: Animal Consciousness, Animal Pain, and Science.* New York: Oxford University Press.

Rose JD. 2007. Anthropomorphism and "mental welfare" of fishes. *Diseases of Aquatic Organisms* 75: 139–154.

Roughgarden J. 2004. *Evolution's Rainbow: Diversity, Gender, and Sexuality in Nature and People.* Berkeley: University of California Press.

Sapolsky RM. 2001. *A Primate's Memoir: A Neuroscientist's Unconventional Life among the Baboons.* New York: Touchstone.

Schaik CP van, Ancrenaz M, Borgen G, Galdikas B, Knott CD, Singleton I, Suzuki A, Utami SS, Merrill M. 2003. Orangutan cultures and the evolution of material culture. *Science* 299: 102–105.

Schusterman RJ, Gisiner RG, Hanggi EB. 1993. Remembering in California sea lions: Using priming cues to facilitate language-like performance. *Animal Learning and Behavior* 21: 377–383.

Silverman J. 2009. Sentience and sensation. *Lab Animal* 37: 465–467.

Sinn DL, Perrin NA, Anderson RC, Mather JA. 2001. Early temperamental traits in an octopus (*Octopus bimaculoides*). *Journal of Comparative Psychology* 115(4): 351–364.

Skelhorn J. 2008. Comment on Murdochs, The last word: Foxy tale. *New Scientist* 2677: 81.

Slobodchikoff CN, Kiriazis J, Fischer C, Creef E. 1991. Semantic information distinguishing individual predators in the alarm calls of Gunnison's prairie dogs. *Animal Behaviour* 42: 713–719.

Sneddon LU. 2003. The evidence for pain in fish: The use of morphine as an analgesic. *Applied Animal Behaviour Science* 83: 153–162.

———. 2009. Pain perception in fish: Indicators and endpoints. *ILAR Journal* 50: 338–342.

Steinfeld H, Gerber P, Wassenaar T, Castel V, Rosales M, Haan C de. 2006. *Livestock's Long Shadow: Environmental Issues and Options.* Rome: Food and Agriculture Organization.

Strieck F. 1924. *Sensory, neural, and behavioral physiology.* Z. vergleich Physiol. 2: 122–154.

Struhsaker TT. 1967. Auditory communication among vervet monkeys (*Cercopithecus aethiops*). In *Social Communication among Primates,* ed. Altmann SA, pp. 281–324. Chicago: University of Chicago Press.

Tate AJ, Fischer H, Leigh AE, Kendrick KM. 2006. Behavioural and neurophysiological evidence for face identity and face emotion processing in animals. *Philosophical Transactions of the Royal Society B* 361: 2155–2172.

Tebbich S, Bshary R, Grutter A. 2002. Cleaner fish (*Labroides dimidiatus*) recognise familiar clients. *Animal Cognition* 5: 139–145.

Teyke T. 1985. Collision with and avoidance of obstacles by blind cave fish *Anoptichthys jordani* (Characidae). *Journal of Comparative Physiology A* 157: 837–843.

———. 1989. Learning and remembering the environment in the blind cave fish *Anoptichthys jordani*. *Journal of Comparative Physiology A* 164: 665–662.

Thornhill R, Gangestad SW. 1999. Facial attractiveness. *Trends in Cognitive Science* 3: 452–460.

Tolimieri N, Haine O, Jeffs A, McCauley R, Montgomery J. 2004. Directional orientation of pomacentrid larvae to ambient reef sound. *Coral Reefs* 23: 184–191.

Uexküll J von. 1934. A stroll through the worlds of animals and men: A picture book of invisible worlds. In *Instinctive Behavior: The Development of a Modern Concept,* ed. and trans. Schiller CH, pp. 5–80. New York: International Universities Press, 1957.

Uhlenbroek C. 2002. *Talking with Animals.* London: Hodder and Stoughton.

de Waal F. 2001. *The Ape and the Sushi Master.* New York: Basic Books.

———. 2005. *Our Inner Ape.* New York: Riverhead Books.

de Waal F, Lanting F. 1997. *Bonobo: The Forgotten Ape.* Berkeley: University of California Press.

Watson DM, Croft DB. 1996. Age-related differences in play fighting strategies of captive male red-necked wallabies (*Macropus rufogriseus banksianus*). *Ethology* 102: 336–346.

Weatherhead PJ, Robertson RJ. 1979. Offspring quality and the polygyny threshold: "The sexy son hypothesis." *American Naturalist* 113: 201–208.

Wilkinson GS. 1992. Communal nursing in the evening bat *Nycticeius humeralis. Behavioral Ecology and Sociobiology* 31: 225–235.

Winfrey C. 2008. Standing tall: Niger's giraffes and our 16th president. *Smithsonian* 39: 8.

Winkler E. 2009. A short diversion into the sex life of animals. *Equal Writes* (blog), January 28. Accessed August 22, 2010. http://equalwrites.org/2009/01/28/a-short-diversion-into-the-sex-life-of-animals.

Winterbottom M, Burke T, Birkhead TR. 2001. The phalloid organ, orgasm and sperm competition in a polygynandrous bird: The red-billed buffalo weaver (*Bubalornis niger*). *Behavioral Ecology and Sociobiology* 50: 474–482.

Yasuoka A, Abe K. 2009. Gustation in fish: Search for prototype of taste perception. *Results and Problems in Cell Differentiation* 47: 239–255.

Young R. 2003. *The Secret Lives of Cows.* Preston, U.K.: Farming Books and Videos.

Zuk M. 2003. *Sexual Selections: What We Can and Can't Learn about Sex from Animals.* Berkeley: University of California Press.

INDEX

adaptive behavior: play as, 25, 36; pleasure and, 3, 6, 68–69; sex as, 88; thrill seeking as, 171

adaptive traits, beauty and appearance as, 170, 171

aesthetic beauty and appreciation, 13, 169–71, 175

affection, 68, *147. See also* love and emotional attachment

African elephant, *23–24*, 108–9, *193*

African lion, *126*

alarm calls, 2, 11, 36, 148

alcohol: animal intoxication, 171–72

alliesthesia, 9, 127

allogrooming. *See* grooming behaviors

allopreening, *66*, 68, *72–73*, *80*, *111*, *149*

Alpine ibex, *36*

altruism. *See* empathy and consideration

Amblyrhynchus cristatus (marine iguana), 68–69, *71*

American black bear, 192, *194–95*

American pika, *57*

American robin, *19*

anecdotes, vs. scientific studies, 2

angelfish, six-barred, *15*

animal agriculture, 186–87; legal regulation of, 187–88

animal cognition, 4, 5, 13–14

animal existence and experience, 2–3, 4, 188–89; human prejudices about, 10–11, 107–8, 192. *See also* feelings; sentience

animal-human relationships, 185–92

animal protection laws, 187–88

animal sanctuaries, 57, 67, 163, 176, 191

Animals Asia Foundation, 176

animal welfare, 185–86, 187–92

anthropodenial, 4

anthropomorphism, 3–4, 108

Antidorcas marsupialis (springbok), *180*

ants, 12–13

apes, 45, 68, 108, 173. *See also specific apes*

appearance, aesthetic appreciation and, 170–71

Arabian babbler, 26

archerfish, *49*

Arctic ground squirrel, *141*

Ardea herodias (great blue heron), 128, *130–31*

Artamus cyanopterus (dusky woodswallow), *149*, *154*

Asian black bear, *177*

Asiatic lion, *34–35*, *116*

Athene noctua (little owl), *81*

Atlantic gray seal, *113*, 188–89, *188*

autoeroticism, 88, 89

awareness, 2–3

babbler, 26, *54–55*, *80*

baboon, 11, 26, 46, 89, 107, 149–50

Baboon Matters, 26

Baeolophus bicolor (tufted titmouse), 128, *136*

Bagemihl, Bruce, 88–89

Barbary macaque, *78*, *151*, *183*

barbel, 48, 70

barnacle goose, *94–95*

barter, 87–88

basking, 17–18, 128, *136*, *140*

bats, 28, 109, 172, 186; food and foraging, 46, 47, *52–53*, 172

bears, 27, 30, *135*, 177, 192, *194–95*

beauty. *See* aesthetic beauty and appreciation

beaver, *112–13*

bee-eater, carmine, *156*

beetles, *8*, 58, 170, 172; yellow weevil, 170–71, *174–75*

behavior: alternative explanations for, 4–5; as rewarding, 10–11

Bekoff, Marc, 4

Belding's ground squirrel, *37*

beluga, *24*

Bengal cat, *124–25*

billing, *121*

Biological Exuberance: Animal Homosexuality and Natural Diversity (Bagemihl), 88–89

biomimicry, 171

birds: alarm calls, 11, 148; allopreening, *66, 68, 72–73, 80, 111,* 149; comfort behaviors, *128, 129, 130–31, 136–37, 141;* companion interactions, *147–48, 149, 154, 155, 156;* courtship behaviors, *60–61, 92–93,* 109, 110, *118–19, 121,* 156, 170; curiosity, 172, *178;* emotional attachments, 73, 107, 109, 110, 121; feeding behaviors, *46–47, 54–55, 58, 59, 60–61;* flight, 1, 4–5, *166–67,* 171, *178;* intoxication, 172; play, 1, 4–5, 26; sexual behavior, 88, *94–95,* 101. *See also specific birds*

black bear, 192, *194–95*

black-headed grosbeak, *121*

black oystercatcher, *104–5*

black redstart, 149, *155*

black-tailed gnatcatcher, *128*

bluegill sunfish, 14

blue jay, 46, *128*

blue-streaked cleaner wrasse, *15, 75*

Bombycilla cedrorum (cedar waxwing), *58, 172*

bonobo, *38–39,* 48, 87–88, 173, *193*

Bos taurus. See cattle

The Botany of Desire (Pollan), 169

bottlenose dolphin, *182–83*

brain chemistry, 7, 45, 109–10, 190

brain structure, 7, 108

Branta leucopsis (barnacle goose), *94–95*

brown bullhead, 48

Bshary, Redouan, 70

Bubulcus ibis (cattle egret), *193*

bullhead, brown, 48

Burgdorf, Jeffrey, 10, 173

Burghardt, Gordon, 4, 69

Burgundy snail, 97

butterflies, *17, 18,* 58, *100, 175*

Cabanac, Michel, 9, 127

California quail (*Callipepla californica), 141*

calls and call interactions, 2, 11, 36, 148, 172. *See also* songs and singing

Camargue horse, *68*

Campylopterus hemileucurus (violet sabrewing), *42–43*

Capra aegagrus (domestic goat), *138–39, 161*

Capra ibex (Alpine ibex), *36*

capuchin monkey, 27, 28, 169, 172

carmine bee-eater, 149, *156*

carnivory (human), 186–87

carp, 48

carrion crow, *136*

Castor canadensis (beaver), *112–13*

cats, domestic: comfort behaviors, *124–25,* 129; companionship, 148–49, *161;* individual food preferences, 47; play, 25, *33;* pleasure in touch, 1–2, *68*

cattle, *12, 26–27, 146,* 149, *176, 191*

cattle egret, *193*

cave fish, 14

cedar waxwing, *58, 172*

cephalopods, 13; octopuses, 8, 13

Cervus canadensis (elk), *16–17, 16, 181*

Cervus elaphus (red deer), *98–99*

chasing behaviors, 1, 4–5, 35

cheating, 70

cheetah, *181*

Chen caerulescens (snow goose), *166–67*

chickadee, 148

chickens, 8, 11–12, 148, 169–70, 191

child-parent bonds. *See* parent-child bonds

chimpanzee, 27, 48, 150, 173; companionship, *162–63;* emotional attachments, 109, 110

Chimp Haven, 163

cichlids, 70

cleaner fish, 9, 14, *15,* 69–70, 74

climate change, 186–87

Coccinellidae ladybird beetle, *8*

cockatoo, sulphur-crested, 47

comfort, 127–29; photographs, *124–41*

common blue butterfly, *100*

common chimpanzee. *See* chimpanzee

common marmoset, 110

common porcupine, *44*

common tern, *60–61*

companionship, 15, 129, 147–50; photographs, *144–63*

consciousness, 2–3

consideration. *See* empathy and consideration

contrafreeloading, 11

cooperation, 28, 148

copulation. *See* mating and sexual behavior

Corvus corone (carrion crow), 136

Corvus corone cornix (hooded crow), 31

courtship behaviors, 109; of birds, 60–61, 92–93, 109, 110, 118–19, 121, 156, 170; of butterflies, 101; of sharks, 90, 97; of turtles, 18

Cowper, William, 192

cows. *See* cattle

crayfish, 8

crows, 1, 4–5, 26, 31, 136

Cummings, Terry, 67

Curculionidae weevil, 170–71, 174–75

curiosity, 172–73; photographs, 168, 178, 181

cutthroat trout, 189, 190

Cynomys ludovicianus (prairie dog), 115

Darwin, Charles, 4, 5, 107

deception, 11–12

Deeble, Mark, 70

deer, 64–65, 98–99, 148–49, 161

Delphinapterus leucas (beluga), 24

Dendrocopos major (great spotted woodpecker), 59

diagonal-banded sweetlips, 144–45, 147

dikdik, 11

dogs, 2, 4, 77, 183; feral dog-langur monkey interactions, 70, 77; individual food preferences, 47; laughter, 173; legal protections for, 188; play, 25, 27; pleasure in touch, 1, 68; sense of fairness, 27–28

dolphins, 16, 25, 86, 182–83, 186; consideration for others, 28, 147

domesticated animals. *See* farmed animals; *specific animals*

dominance displays, 4–5

dopamine, 7, 109, 110

Douglas-Hamilton, Iain, 198

dusky woodswallow, 149, 154

dustbathing, 141

eastern gray squirrel, 157

egret, 193

Egyptian fruit bat, 52–53

Eisley, Loren, 192

elephants, 23–24, 108–9, 128–29, 193; empathy, 28, 108–9; intoxication, 172

elk, 16–17, 16, 181

emotional attachment. *See* companionship; love

emotions. *See* feelings; *specific emotions*

empathy and consideration: among animals, 28, 107, 108–9, 147, 148, 150; human empathy for animals, 188–89

Encyclopedia of Animal Behavior, 107

Enhydra lutra (sea otter), 15, 158

Equus ferus (konik polski horse), 78–79

Erethizon dorsatum (common porcupine), 44

European brown frog, 84–85

European Union animal protection laws, 188

experience, 2, 18

The Expression of the Emotions in Man and Animals (Darwin), 4, 5, 107

facial expressions, 16, 45, 127

facial recognition skills, 12

fairness, sense of, 27–28

familial love. *See* love; parent-child bonds

farmed animals, 11–12; climate change and animal agriculture, 186–87; legal protections for, 187–88; sanctuaries, 57, 67, 176, 191. *See also specific animals*

Farm Sanctuary, 57, 176

feelings, 3; biological bases of, 7, 12, 108; intensity of, 10; physical expressions of, 16, 45, 68, 69, 107, 149–50. *See also* sentience; *specific emotions*

Felis catus. *See* cats, domestic

Felis onca (jaguar), 98

fin whale, 108

"Fish" (Lawrence), 69

Fish Cognition and Behavior, 13

fishes, 13–14, 189–90; cleaner fishes, 9, 14, 15, 69–70, 74, 75; companionship and sociability, 149, 155, 190; hermaphroditism, 89; play, 26; sensory perception and responses, 48, 172–73, 189. *See also specific fishes*

flight, 166–67, 171, 179; aerial chases, 1, 4–5

flying fox, gray-headed, 32

food and foraging, 11, 45–48; ants, 12; bats, 46, 47, 52–53, 172; bluegill sunfish, 14; courtship feeding, 60–61, 156; feeding as social behavior, 48, 147–48; group hunting and foraging, 28, 147–48; individual food preferences, 47–48; photographs, 42–61; sharing food, 28, 121

food chains, 186–87

foxes, 26–27, 37, 180

friendship, 149. *See also* companionship

frigatebird, 5

frill-fin goby, 14
frogs, *84–85, 87*
fruit bat, 46, *52–53*

Gaia, Veronika, 17
games. *See* play
geese, *94–95,* 109, 147–48, *166–67*
gender conversion, 89
gender roles and differences, 89, 99, 171
gentoo penguin, 110, *111*
GG (genito-genital) rubbing, 88
Giraffa camelopardalis. See giraffes
Giraffa camelopardalis tippelskirchi (Masai giraffe), *91*
giraffes, 68, 91, 117
global warming, 186–87
goats: domestic goat, *138–39, 161;* ibex, *36;* mountain goat, *160*
gobies, 14, 89
Goodall, Jane, 109
The Good Good Pig (Montgomery), 129
Gore, Al, 187
gorillas, *114,* 173
gray-headed flying fox, *32*
gray langur, *29, 52, 77*
gray squirrel, *134, 157,* 171
great blue heron, 128, *130–31*
great spotted woodpecker, *59*
grebe, pied-billed, 109, *118–19*
greeting behavior, *115*
grief, 107, 108–9, 149–50
Griffin, Donald, 4
griffon vulture, *66*
grizzly bear, *135*
grooming behaviors, 68; beavers, 113; birds, 66, 68, 72–73, 80, *111,* 149; cleaner fishes, 9, 14, 15, 69–70, *74, 75;* dogs and langur monkeys, 77; primates, 68, *77, 78,* 87–88, 150, *151, 162–63*
grosbeak, *121*
ground squirrel, *37, 141*
gulls, 5, *31*
guppy, 149
Gyps fulvus (griffon vulture), 66

Haematopus bachmani (black oystercatcher), *104–5*
hairy woodpecker, 46
Halichoerus grypus (Atlantic gray seal), *113,* 188–89
hamlet, 89
Heinrich, Bernd, 109

Helix pomatia (Burgundy snail), *97*
Hermanson, Becky, 148–49
hermaphroditism, 89, 97
heron, great blue, 128, *130–31*
herring gull, *31*
hippopotamuses, 9, 70, 74
hoary marmot, *33, 50–51*
Hoerauf, Dave, 67
Hogwood, Christopher, 129
homeostasis, 127
homosexuality, 88–89
hooded crow, *31*
horses, 68, 78–79
house sparrow, *101*
Hoyle, Rich, 27
huddling, 128, 155
human-animal relationships, 185–92
Humane Society of the United States, 187
human pleasure, 6
hummingbirds, *42–43*
humor, 173
humpback whale, 108
hyena, 147

ibex, *36*
iguana, 9, 46, 68–69, 71
individuality, 192
inequity aversion, 27–28
information sharing, 147; through calls, 2, 11, 36, 148, 172
insects, *8,* 46–47, 170–71, *174–75. See also specific insects*
instinct, 89
intelligence, 186
Intergovernmental Panel on Climate Change (IPCC) report (2007), 186, 187
intoxication, 171–72
intrinsic value, 19
invertebrates, *8, 12–13,* 46–47. *See also specific invertebrates*

jackdaw, 28
jaguar, *98*
Japanese macaque, *76–77,* 106, *133*
Jersey Wildlife Preservation Trust, 47
joy, 2, 9
jungle babbler, *54–55, 80*
justice, 28

Kallima inachus (orange oakleaf butterfly), *175*
kea, *178*
kelping, 75
Kendrick, Keith, 12
killer whale. *See* orca
killing animals, 185
"kissing," *115, 121*
Koko, 173
konik polski horse, *78–79*

labeo, 70, 74
Labeo cylindricus (red-eyed labeo), *74*
Labroides dimidiatus (blue-streaked cleaner wrasse), *15, 75*
ladybird beetle, *8*
land snail, *89, 97*
langur monkey, *29, 52, 70, 77*
Larus argentatus (herring gull), *31*
laughter, 173
Lawrence, D. H., 68–69
laws, animal protection, 187-88
learning capacity, 48, 172. *See also* memory
leisure time, 11
lemur, *132–33, 140, 148*, 172
Lemur catta (ring-tailed lemur), *140*
leopard shark, *96, 97*
lion, *34–35, 116, 126, 147*
little owl, *81*
longtail weasel, *168*
Lorenz, Konrad, 109
lorikeet, *47, 120–21*
love and emotional attachment, 15, 107–11; birds, 73, 107, 109, 110, 121; pair bonds, 68, 94, 107, 109, 110–11; photographs, *104–21. See also* parent-child bonds
Loxodonta africana (African elephant), 23–24, *108–9, 193*
Lycaenidae butterflies, *58*

macaques, 46; Barbary macaque (*Macaca sylvanus*), *78, 151, 183*; Japanese macaque (*Macaca fuscata*), *76–77, 106, 133*
macaw, *72–73*
magpie, *26*, 30
mammals, 68, 128. *See also specific mammals*
map pufferfish, 75
Margulis, Jennifer, 68
marine iguana, 68–69, *71*
marmoset, 110
marmot, hoary (*Marmota caligata*), *33, 50–51*

Masai giraffe, *91*
massages, 67–68. *See also* touch
masturbation, 88, 89
maternal love, 110, 115, 116. *See also* parent-child bonds
Mather, Jennifer, 13
mating and sexual behavior, 18, 87–90, 170, 171; pair bonds, 68, 94, 107, 109, 110–11; photographs, *84–101. See also* courtship behaviors
McGregor, Sheila, 26–27
meat eating (by humans), 186–87
meerkat, *128*, 148
memory, 2–3, 13–14, 45, 48, 148, 190
Merops nubicoides (carmine bee-eater), *156*
Mexican cave fish, 14
mice, 68, *128*, 150, 191
mimicry, 171
minnow, 48, 149
mobility, 6
mollusks, 13. *See also specific mollusks*
mongoose, 70
monkeys: alarm calls, 2, 148; female orgasm, 88; grooming behaviors, 68, 107; intoxication, 172; preference for symmetry, 169; taste responses and food preferences, 45–46. *See also specific monkeys*
Montgomery, Sy, 129
moon bear, *177*
morality: ethics of human-animal relationships, 185–86, 191–92; play and, 28
Moss, Cynthia, 198
mountain goat, *160*
mule deer, 148–49, *161*
Mustela frenata (longtail weasel), *168*
mutualisms, involving touch, 69–70; cleaner fishes, *9*, 14, *15*, 69–70, *74*, 75. *See also* grooming behaviors
Mzima Springs, 70

naked mole rat, 28
narrow-snouted barbel, 70
Naskrecki, Piotr, 171
nature: human relationship to, 186; portrayals of, 10
Nestor notabilis (kea), *178*
Nielsen, Anders, 17–18
nociception, 14
norepinephrine, 109
northern hawk owl, 18–19, *18*
Norway rat, *77, 159. See also* rats

Ochotona princeps (American pika), 57
octopuses, 8, 13
Odocoileus hemionus (mule deer), 148–49, *161*
Odocoileus virginianus (white-tailed deer), *64–65*
olfaction, 48
orange oakleaf butterfly, *175*
orangutans, 171, 173
orca *(Orcinus orca), 74,* 108
Oreamnos americanus (mountain goat), *160*
orgasm, 88, 111
osprey, *179*
owls, 18, 19, *81*
Oxford Companion to Animal Behavior, 107
oxpecker, 148
oxytocin, 7, 110–11

pain, 6, 7–8, 14, 150, 190
pair bonds, 68, 94, 107, 109, 110–11
pallid bat, 47
Pandion haliaetus (osprey), *179*
Panksepp, Jaak, 10, 173
Pan paniscus (bonobo), *38–39,* 48, 87–88, 173, *193*
Panthera leo (African lion), *126*
Panthera leo persica (Asiatic lion), *34–35,* 116
Pan troglodytes. See chimpanzee
parakeet, 172
parent-child bonds: apes and monkeys, 107, 109, 115, *133;* other mammals, 116, *117, 132–33, 194–95*
parrot, 47, 68; macaw, *72–73*
Passer domesticus (house sparrow), *101*
Pavlov, Ivan, 4
penguin, 68, 110, *111*
pheasant, ring-necked *(Phasianus colchicus), 92–93*
Phenicurus ochruros (black redstart), 149, *155*
Pheucticus melanocephalus (black-headed grosbeak), *121*
pied-billed grebe, 109, *118–19*
pigeon, 128
Pig Preserve, 27
pigs, 27, *56,* 129, *152–53,* 191; wild boar, 116
pigtailed macaque, 46
pika, 57
plants, 12; aesthetic attractions of, 169
play, 1, 3, 13, 25–28, 148; laughter and, 173; photographs, *23–39, 182–83;* stotting or pronking, *180;* thrill seeking, 171
play bowing, 16, 27
play fighting, 27, 30, 32

Pleasurable Kingdom: Animals and the Nature of Feeling Good (Balcombe), 19
pleasure, 1–2, 3, 5–9, 191–92; as adaptive, 6, 68–69; alliesthesia, 9, 127; biochemistry of, 7, 45, 109–10; biological bases of, 7; evolution of, 6; food choices and, 45, 46; as motive for sex, 88–89; physical/behavioral expressions of, 16, 45, 68, 76, 110; scientific neglect of, 2, 3, 87; survival behaviors and, 10–11, 88
Plectorhinchus lineatus (diagonal-banded sweetlips), *144–45, 147*
Podilymbus podiceps (pied-billed grebe), 109, *118–19*
Podocnemis unifilis (yellow-spotted side-necked turtle), 17–18, *17*
polar bear, 27, 30
Pollan, Michael, 169
Polyommatinae butterfly, 58
Polyommatus icarus (common blue butterfly), *100*
Pomacanthus sexstriatus (six-barred angelfish), *15*
Poole, Joyce, 198
Poplar Spring Animal Sanctuary, 67, 191
porcupine, common, *44*
post-traumatic stress disorder, 108
prairie dog, *115,* 129, 148
prairie vole, 110–11
prawn, 8
preening, 68; allopreening, 66, 68, *72–73,* 80, 111, 149
prejudice, 11, 185
primates: feeding behaviors, 48; grooming behaviors, 68, 79, 87–88, 107, 150, *151, 162–63;* play, 26; taste responses, 45. *See also specific primates*
pronking, 181
Propithecus candidus (silky sifaka), *132–33*
Pteropus poliocephalus (gray-headed flying fox), *32*
PTSD (post-traumatic stress disorder), 108
Pygoscelis papua (gentoo penguin), 110, *111*

quail, *141,* 148
The Question of Animal Awareness (Griffin), 4

rail, Virginia, *136–37*
rainbow lorikeet, 47, *120–21*
Rallus limicola (Virginia rail), *136–37*
Rana temporaria (European brown frog), *84–85*
rats, 7–8, 9, 77, 129, *159;* empathy, 28; food preferences and taste responses, 45, 46, 47; laughter, 173; touch and grooming behaviors, 68, 76

Rattus norvegicus (Norway rat), 77, *159*

raven, 26, 30, 109

recognition skills: fish, 2, 10, 155, 190; other animals, 11, 12, 13, 32, 149

red deer, *98–99*

red-eyed labeo, *74*

red fox, *37*

red-fronted macaw, *72–73*

red squirrel, *35*

redstart, black, 149, *155*

reptiles, 26, 46, 127–28; iguanas, 46, 68–69, *71*

ring-necked pheasant, *92–93*

ring-tailed lemur, *140*

risky behavior, 171–72

robin, American, *19*

Robinson, Jill, 176

rodents, 32, 57, 68. *See also specific rodents*

Rollin, Bernard, 10

roosters, 11–12

Rose, James, 14

Rousettus aegyptiacus (Egyptian fruit bat), *52–53*

sac-winged bat, 109

safety and security, 147–48

Salmo clarki (cutthroat trout), 189, *190*

sandfish, 89

sandpiper, 129

Sapolsky, Robert, 11

scarlet macaw, *72–73*

Sciurus carolinensis (eastern gray squirrel), *157*

Sciurus griseus (western gray squirrel), *134*

scratching, 128–29, *130–31, 138–39*

sea bass, 89

seal, Atlantic gray, *113,* 188–89, *188*

sea otter, 15, *158*

Semnopithecus entellus (northern plains gray langur), *29, 52, 77*

sensory perception and experiences, 6–7, 10, 12, 189; sexual pleasure, 88; taste sensitivity and olfaction, 45–46, 47–48

sentience, 3, 6, 9, 18–19, 192; fishes, 13–14, 189–90; human prejudices and controversy about, 3–5, 10, 185–86, 189; invertebrates, 8, 12–13. *See also* feelings; sensory perception

sentries, 147–48

sequential hermaphroditism, 89

serotonin, 7, 109

sex. *See* mating and sexual behavior

"The Sex Lives of Animals" (museum exhibit), 89

sexy son hypothesis, 170

sharing, 28

sharks, 89–90, *96, 97*

Shaw, George Bernard, 48

sheep, 12, 67–68

Siamese fighting fish, 14

silky sifaka, *132–33*

Simonet, Patricia, 173

simultaneous hermaphroditism, 89, 97

six-barred angelfish, *15*

sleep, 15, 108, 127, *135, 158*

The Smaller Majority (Naskrecki), 171

snails, 89, 97

snowfinch, 57

snow goose, *166–67*

social behavior: comfort behaviors as, 128, 129; feeding as, 48; touch as, 68. *See also* companionship; grooming behaviors; mutualisms

solipsism, 9

songs and singing: bats, 109, 170; birds, 110, *121.* *See also* calls and call interactions

sour tastes, responses to, 45–46

sparrow, house, *101*

Spermophilus parryii (Arctic ground squirrel), *141*

sperm whale, 108

spider monkey, 45, 46

spindle cells, 108

spinner dolphin, *86*

springbok, *180*

squirrel monkey, 46, 169

squirrels: companionship and comfort behaviors, *134, 141, 157;* play, 35, 37; risky behaviors, 171

starling, 47, 128, 149

Stegostoma fasciatum (leopard shark), *96, 97*

Stenella longirostris, 86

Sterna hirundo (common tern), *60–61*

stickleback, 14

Stone, Vicky, 7

stotting, 181

stretching, 127, 129, *137, 140*

substance abuse, 171–72

suckers, 48

sulphur-crested cockatoo, 47

Surnia ulula (northern hawk owl), *18, 19*

survival behaviors: need to engage in, 11, 57; pleasure and, 10–11, 88

Sus scrofa (wild boar), *116*

Sus scrofa domesticus (domestic pig). *See* pigs
sweetlips, diagonal-banded, *144–45, 147*
sweet tastes, responses to, 45, 46
swift fox, *180*
symmetry, 169–70

tamarin, 148
Tamiasciurus hudsonicus (red squirrel), *35*
"The Task" (Cowper), 192
taste sensitivity and food preferences, 45–46,
    47–48
teasing, 26, 30
tern, common, *60–61*
Thalarctos maritimus (polar bear), 27, 30
thermoregulating behaviors, 17–18, 128;
    photographs, *135, 136, 140, 155, 157*
thermoregulation, 127–28
threat displays, 14
thrill seeking, 171–72
tickling, 173
titmouse, tufted, 128, *136*
toads, 87
touch, 67–70, 110, 111; cleaner fishes, 9, 14, *15,*
    69–70, 74, 75; photographs, *64–81;* sea otters, *15;*
    sharks, 90. *See also* grooming behaviors
Toxotes jaculatrix (archerfish), *49*
Trichoglossus haematodus (rainbow lorikeet), *47,*
    120–21
trout, 14, 189, *190*
tufted titmouse, 128, *136*
Turdoides striatus (jungle babbler), *54–55, 80*
Turdus migratorius (American robin), *19*
Tursiops truncatus (bottlenose dolphin), *182–83*
turtle, yellow-spotted side-necked, 17–18, *17*

Uexküll, Jakob von, 10
Umwelt, 10
The Unheeded Cry (Rollin), 10

Urocitellus beldingi (Belding's ground squirrel), *37*
Ursus americanus (black bear), 192, *194–95*
Ursus horribilis (grizzly bear), *135*
Ursus thibetanus (Asian black bear), *177*

vasopressin, 110
vervet monkey, 2
vigilance, 147–48
violet sabrewing, *42–43*
Virginia rail, *136–37*
vole, 110–11
Vulpes velox (swift fox), *180*
Vulpes vulpes (red fox), *37*
vulture, 66, 149

Waal, Frans de, 4, 87
wake surfing, *182–83*
Walsberg, Glenn E., 128
warbler, 148
warthog, 70
weasel, longtail, *168*
weaver, red-billed buffalo, 88
weevil, 170–71, *174–75*
western gray squirrel, *134*
western lowland gorilla, *114*
whales, 24, 74, 108, 147
white-tailed deer, *64–65*
wild boar, *116*
wolves, 147
woodpecker, 46–47, 59
woodswallow, 149, *154*

yellow-spotted side-necked turtle, 17–18, *17*

zebra finch, 110
zebras, 79
zebra shark, 90
Zuk, Marlene, 46, 88